# The Phoenix Strain

A Thriller of Genetic Destiny and Human Survival

## Donald J. Wright

# Contents

# Chapter 1: The Banality of Evil

D r. Ben Carter had been counting exits since the moment he walked in. Four doors. Two stairwells. One service corridor behind the lectern.

Old habits. Habits he couldn't kill.

He rested his hand lightly on the podium and studied the faces watching him. Forty-three students stared back—he'd counted them as they filed in. He'd tallied them as they filed in, scanning for anomalies. The university was supposed to be a safe environment. Safety was an illusion.

He tapped the clicker. An image flashed onto the screen—a sunlit villa beside a frozen lake. Idyllic. Deceptive.

The morning light streaming through the tall Gothic windows of Berlin's Friedrich-Wilhelm University caught the dust motes dancing in the air. Somewhere in the back row, a student was trying to discreetly eat what appeared to be a breakfast pretzel.

Three years of teaching, and he still catalogued details like a threat assessment: exits, sight lines, and behavioral anomalies. The academic world felt safe, but safety was often an illusion.

Three years of teaching, and he still catalogued details like a threat assessment: exits, sight lines, and behavioral anomalies. The academic world felt safe, but safety was often an illusion.

"The Wannsee Conference," Ben began, his voice carrying easily across the tiered seating, warm and conversational despite the chill of the subject matter, "lasted exactly ninety minutes. Ninety minutes to systematize the murder of eleven million human beings." He clicked to the next slide, revealing a photograph of an elegant lakeside villa that looked deceptively peaceful in the winter sunlight. "The villa still stands. You can take the S-Bahn there this afternoon if you like. Have coffee in the same room where they planned genocide."

He paused, watching the pretzel-eater in the back row slowly lower his breakfast. Good. Sometimes the proximity of evil—that it had happened here, in their city, in places they passed every day—could penetrate even the most distracted minds.

"The minutes of the meeting were clinical and efficient," Ben continued, moving away from the podium toward the students. It was a habit that drew them into conversation rather than keeping them at arm's length. "No theatrical speeches about racial superiority. No ranting about the master race. Just logistics." His gray eyes swept the room, taking in the subtle shift in posture as the students leaned forward. "How to transport them. How to process them. How to dispose of them. Evil rarely announces itself with fanfare, ladies and gentlemen. It fills out forms in triplicate and submits them by the deadline."

Anna Weber, a third-year student with purple-streaked hair and the kind of earnest intensity that reminded him why he'd chosen teaching over the shadowy world of intelligence analysis, raised her hand from the middle section.

"Professor Carter, how could ordinary bureaucrats participate in something so monstrous? I mean, these weren't SS storm troopers or concentration camp guards. They were educated men with families, right? University degrees, careers..." She hesitated, then pressed on. "Weren't they just following orders?"

The question hung in the air like smoke from a distant fire. Ben had answered variations of it hundreds of times over the past three years, but today it felt different. Heavier. The faces in the room seemed to blur for a moment, replaced by other faces—men in expensive suits around a mahogany table, discussing GPS coordinates and acceptable casualty rates. His throat tightened almost imperceptibly.

"Following orders," he repeated slowly, then shook his head with a rueful smile that didn't quite reach his eyes. "Anna, that's perhaps the most dangerous phrase in any language." He moved to the far end of the platform, closer to the students, his voice dropping slightly to draw them in. "Hannah Arendt wrote about the banality of evil. These weren't monsters, not in the way we'd like to imagine. They were clerks, administrators, and middle managers. They had university degrees, season tickets to the opera, and family photographs on their office desks."

He stopped beside a window, the winter light from the Berlin street below casting long shadows across the lecture hall. The city hummed with its usual morning energy—trams clanging, construction sounds, the distant laughter of students hurrying to class. Life persisted, as it always did, even in the shadow of history's darkest chapters.

"They went home to their families every evening," Ben continued, his voice carrying a note of genuine sorrow that made several students lean forward unconsciously. "They played with their children, worried about their mortgages, and complained about the weather. The horror isn't that they were inhuman. The horror is that they were entirely human. And when the time came to choose between career advancement and moral courage, between following orders and taking a stand..." He turned back to face the room, his expression grave. "Well. History shows us what they chose."

Marcus, a quiet student from Munich who rarely spoke up, tentatively raised his hand. "But surely someone must have objected? Spoken up?"

Ben's expression softened slightly. These students needed to believe in the possibility of resistance, of individual courage making a difference. He'd once believed in it too, before he'd learned how easily good intentions could be corrupted by circumstance and necessity.

"One person did question the proposed methods," Ben said carefully. "Not the morality—the efficiency. He was concerned about the psychological impact on the executioners, not the victims. Even conscience, when it appeared, was filtered through the system it served."

The silence that followed was pregnant with an uncomfortable truth. Ben could hear the familiar sounds of the university—footsteps in the corridor, muffled conversations, the distant hum of the heating system. Normal sounds of a normal Tuesday morning in a normal world, where such horrors were safely confined to history books.

His phone buzzed in his jacket pocket. A brief vibration, barely noticeable, but his hand moved instinctively to silence it. The motion was automatic, practiced—muscle memory from another life, where an unexpected communication could mean the difference between life and death.

"The most chilling document from that meeting," Ben continued, his voice steady despite the slight tremor in his chest, "isn't the official minutes. It's a handwritten note found in the margin of one attendee's copy. 'Remember to pick up milk on the way home.' Genocide and groceries, planned in the same ninety minutes."

Several students shifted uncomfortably in their seats. The mundane detail somehow made the horror more real, more immediate. Ben had learned long ago that it was the small human touches that made atrocities comprehensible—and therefore preventable.

"Your assignment for Thursday," he said, beginning to gather his notes, "is to read the Wannsee Protocol in its entirety. Pay attention to the language, the euphemisms. 'Final solution.' 'Resettlement.' 'Special treatment.' Notice how evil disguises itself behind pleasant words, how it makes itself sound not just reasonable, but necessary."

The familiar rustle of papers and closing laptops filled the air as students prepared to leave. Ben watched them file out, their youthful faces already turning toward whatever came next—lunch, other classes, the simple concerns of a generation that had never known real fear. He envied them that innocence, even as he worked to preserve it.

"Professor Carter?"

Anna Weber stood by the podium, her expression troubled. "Are you all right? You seemed... I don't know. Different today. More intense than usual."

Ben managed a genuine smile, the kind that had made him popular with students despite his occasionally distant manner. "Just tired, I'm afraid. Long night grading papers." He paused, studying her concerned face. "Your essay on the Nuremberg Trials was particularly thoughtful, by the way. You raised some excellent points about the legal precedents."

She brightened at the praise, then hesitated. "Professor? Do you think... do you think it could happen again? Something like that?"

The question hit him like a physical blow. Ben looked into her earnest young eyes and saw himself at that age—brilliant, idealistic, convinced that knowledge and good intentions could change the world. Before he'd learned that sometimes the people trying to save the world were the most dangerous of all.

"I think," he said carefully, "that vigilance is the price of freedom. We study history not because it's past, but because it's prologue. The capacity for evil doesn't disappear—it just finds new forms, new justifications."

Anna nodded, seeming satisfied with the answer, and headed for the exit. Ben watched her go, then began shutting down his laptop with practiced efficiency. Another lecture delivered, another day in his carefully constructed exile from the world of action.

But as he packed his papers into his worn leather satchel—a gift from his father when he'd started graduate school, back when the future had seemed bright and uncomplicated—his phone buzzed again. This time, he glanced at the screen.

The message was from an encrypted number he didn't recognize, but the format was familiar. Too familiar.

The Phoenix rises at dawn. GPS: 52.5200° N, 13.4050° E. Come alone. —S

Ben's blood went cold. The coordinates were for a location in central Berlin, and the signature could only mean one thing. Sarah Chen, his former handler, was trying to make contact. Sarah, who had supposedly retired from the CIA three years ago. Sarah, who had promised him that his past was buried, that no one would ever find him in his academic sanctuary.

He deleted the message immediately, but his hands were shaking as he powered down the phone. The familiar weight of paranoia settled on his shoulders like an old, unwelcome coat. He'd thought he was done with encrypted communications and coded messages. He'd thought he was free.

But as he locked his office door and stepped out into the Berlin morning, the winter air sharp against his face, Ben couldn't shake the feeling that his carefully constructed new life was about to crumble. The ghosts of his past weren't just whispering warnings anymore—they were screaming.

The walk to his apartment normally took fifteen minutes through the university district's tree-lined streets. Today, he found himself taking a circuitous route, ducking through side streets and doubling back on himself. Old habits returning with muscle memory precision. His eyes scanned doorways, windows, and parked cars. Looking for watchers, for patterns, for anything out of place.

By the time he reached his building—a modest three-story structure in Charlottenburg—his nerves were singing with tension. The building's entrance looked normal. No one is loitering nearby. No unusual vehicles. But something felt wrong.

His apartment was on the second floor, deliberately chosen for its multiple exit routes. As he climbed the stairs, Ben's hand moved instinctively to his jacket pocket where, in another life, he would have carried a weapon. Now it held only his keys and a worn wallet.

The apartment itself was spartanly furnished—a reflection of his academic salary, his neighbors assumed. But Ben knew the truth. The minimal possessions, the easily portable electronics, the lack of personal photographs or mementos—it was all designed for rapid departure. A life that could be abandoned at a moment's notice.

He went immediately to his bedroom closet, moving aside winter coats to reveal a hidden panel. Behind it was a go-bag he'd hoped never to use again: cash in multiple currencies, three different passports, encrypted USB drives, and a Glock 19 with two spare magazines.

The phone buzzed again. This time, the message was even more direct.

Building under surveillance. Time to go. —S

Ben's training kicked in immediately. He grabbed the go-bag, took one last look around the apartment that had been his sanctuary, and headed for the fire escape. Whatever was coming, Dr. Ben Carter, the mild-mannered professor, was about to disappear.

And whoever had sent that message was about to meet the man he'd tried so hard to leave behind.

The Phoenix was rising, and Ben was terrified to discover what it would burn.

# Chapter 2: The Prometheus Project

D r. Eva Richter's fingers moved like a concert pianist's. The protein structures rotated in the air before her, luminous blue helices twisting against the darkness of the lab.

She spoke softly, as if coaxing the molecules themselves.

"Computer, increase helix stability by twelve percent." She said, watching as the protein structure shifted in response, its bonds tightening with mathematical precision. The changes cascaded through the model in real-time, each adjustment triggering a thousand others until the entire structure achieved a new equilibrium.

The model shimmered. Bonds tightened. Chains realigned in real time.

Eva exhaled, feeling that old thrill—life itself, bent to her will. The elegant choreography of biochemistry.

"Beautiful," she murmured.

The monastery enveloped her, a fortress of ancient stone concealing more modern secrets. Outside, the Bavarian Alps slept under a blanket of snow. Inside, quantum processors thrummed beneath the floor, calculating billions of molecular permutations every second.

Her mother would have loved this place. At twenty-eight, Eva was the youngest lead researcher ever hired by Helix Dynamics. Moments like this—when the elegant complexity of life itself seemed to bend to her will—reminded her why she'd fallen in love with biochemistry in the first place. The dance of molecules, the poetry of genetic code, the infinite possibility contained within each double helix.

"The surrounding laboratory was a marvel of modern technology disguised as medieval architecture. Located in a converted thirteenth-century monastery high in the Bavarian Alps, the facility seamlessly blended ancient stone walls with cutting-edge quantum computers. Gothic windows looked out over equipment that cost more than most countries' entire research budgets, while flying buttresses supported server farms that could process more data in an hour than existed in the world as a whole fifty years ago.

Eva had always found the juxtaposition oddly comforting—old and new, tradition and innovation, working in harmony. Her mother used to say that the best science happened when you honored the past while reaching toward the future. Of course, her mother had said a lot of things before the cancer took her five years ago.

"Working late again, Meine Tochter?"

Eva turned at the sound of her father's voice, her smile automatic and warm. Klaus Richter stood in the doorway, his silver hair perfectly styled despite the late hour, his lab coat pristine white against the dim lighting. At sixty-two, he remained an imposing figure—tall, broad-shouldered, with the kind of commanding presence that had made him legendary in European scientific circles.

"Just running a few more simulations," Eva said, gesturing to the floating molecules. "The protein folding patterns are absolutely fascinating. I think I've found a way to increase cellular penetration by thirty percent while maintaining structural integrity."

Klaus stepped into the lab, his pale blue eyes studying the holographic display with professional interest. Something flickered across his face—pride, perhaps, or satisfaction. "Excellent work. The efficiency gains could be... significant for our applications."

Eva caught the slight pause, the way his voice carried weight beyond the literal meaning of his words. Over the past six months, she'd noticed these subtle shifts in her father's speech—moments where his words seemed to carry implications she couldn't quite grasp. When she'd asked him directly about the project's ultimate goals, he'd been evasive, speaking in generalities about "advancing human health" and "revolutionary treatments."

"Papa," she said carefully, turning to face him fully, "what exactly are we treating with this? I know the protein structures inside and out, and I understand the delivery mechanisms better than anyone. However, I still don't know what condition this is designed to address."

Klaus moved to stand beside her, his hand resting gently on her shoulder in a gesture that had comforted her since childhood. "Eva, you know how competitive this field has become. Corporate espionage, government interference, academic theft—we must be extremely careful about information compartmentalization until we're ready for publication."

It was the same answer he'd given her for months, always delivered with that patient, paternal tone that made her feel foolish for asking. But Eva's scientific mind couldn't help but notice the inconsistencies. The security protocols were excessive even for proprietary pharmaceutical research. The funding seemed unlimited, which was unusual for any legitimate biotech company. And some of the equipment, particularly the bio-containment systems, seemed designed for working with far more dangerous materials than anything she'd been told about.

"I trust you, Papa," she said finally, though the words felt heavier than they should have. "It's just... as the lead researcher, I'd like to understand the full scope of what we're creating."

Klaus squeezed her shoulder, his smile warm but somehow distant. "Soon, Meine Tochter. Once we complete the current phase, you'll understand everything. Your mother would be so proud of what you've accomplished here."

The mention of her mother—dead these past five years from the very cancer that had driven Eva into biochemistry in the first place—sent a familiar pang through her chest. It was a card Klaus played carefully, reminding her of the personal stakes, the lives they could save with their work. The greater good justified the secrecy and isolation.

Before Eva could respond, the lab's intercom crackled to life. The voice that emerged was tinged with barely controlled panic.

"Dr. Richter? We need you in Bio-Lab 3 immediately. There's been an incident."

Eva's blood went cold. Bio-Lab 3 was where they conducted live trials on tissue samples, always under strict Level 4 containment protocols. She'd been there just hours ago, working with Hans Mueller, a junior researcher who'd recently joined the project. Hans was brilliant, enthusiastic, and exactly the kind of passionate scientist she'd once been before the questions began.

Klaus was already moving toward the door, his face grim. "Stay here, Eva. This is protocol—"

"No." She was already pulling off her lab coat, reaching for a fresh containment suit from the wall dispenser. "Hans is my responsibility. If something's gone wrong..."

They ran through the corridors, their footsteps echoing off the ancient stone walls. The monastery's architecture seemed to close in around them, shadows stretching like grasping fingers under the harsh fluorescent lighting. Eva's mind raced through possibilities—contamination, equipment failure, human error. All was dangerous in a Level 4 lab, but manageable if they acted quickly.

The sealed door to Bio-Lab 3 was surrounded by security personnel in full hazmat gear, their faces obscured by reflective visors. Through the reinforced glass, Eva could see Hans Mueller on the floor, his body convulsing violently. Even through the protective barriers, she could hear his screams—high, keening sounds that seemed to come from somewhere deeper than his throat.

"What happened?" Klaus demanded, his voice cutting through the chaos with the authority of a man accustomed to being obeyed.

"Exposure to Sample 7-Alpha," one of the senior technicians reported, his voice muffled by his respirator. "Containment breach during the transfer process. Direct skin contact with approximately 0.3 milliliters of active compound."

Eva pressed closer to the glass, her scientific training warning of human horror. Hans's skin was changing before her eyes, darkening to an unnatural mottled gray. His fingers were clawing at his throat, leaving deep gouges that should have been bleeding but seemed to be sealing themselves almost instantly. The tissue appeared to be regenerating even as it was being damaged.

"My God," she whispered. "What is Sample 7-Alpha? I've never seen anything like this in our research protocols."

Klaus's hand closed over her arm, his grip surprisingly strong. "This is beyond your clearance level, Eva. You need to step back."

But Eva couldn't look away. Hans's convulsions were becoming more violent, his back arching in ways that suggested his spine was adapting to stresses that should have been impossible. His eyes had rolled back, showing only white, but she could see movement beneath the lids—rapid, almost mechanical.

"We need to help him," Eva said, reaching for the emergency containment controls. "Get him to medical, start the standard exposure protocols—"

"No." Klaus's voice was sharp, final. "It's too late for that."

As if summoned by his words, Hans's convulsions suddenly stopped. For a moment, Eva thought he might be recovering, but then she saw his eyes. They were open, staring directly at her through the glass, but there was nothing human left in them. His mouth opened in a silent scream, revealing teeth that seemed to have changed shape, becoming sharper and more predatory.

His skin had taken on a translucent quality, and Eva could see something moving beneath it—patterns of growth that looked almost like circuitry, spreading outward from the point of contact. The tissue was rebuilding itself, but not as it had been. It was becoming something else entirely.

Then, as suddenly as it had begun, it was over. Hans collapsed, his body was still going, completely. The monitors flatlined, their alarms echoing through the sealed chamber. The silence that followed was broken only by the hiss of the ventilation system working overtime to clear the contaminated air.

"Time of death, 22:47," Klaus announced with clinical detachment. "Begin immediate containment and disposal protocols."

Eva stared at him in shock. "Disposal? Papa, we need to understand what happened. We need to analyze the tissue samples and determine how the compound affected cellular structure."

"No analysis," Klaus interrupted, his tone brooking no argument. "This incident is classified at the highest levels. The body will be incinerated within the hour."

The security team was already moving into action, their movements precise and well-rehearsed. Too well-rehearsed, Eva realized with growing horror. This wasn't the first time they'd dealt with this kind of "accident."

"Papa," she said slowly, "what is this project really about? What are we creating here?"

Klaus was already turning away, barking orders to the security team about cleanup procedures and documentation protocols. But Eva caught his arm, forcing him to look at her. For a moment, his mask slipped, and she saw something in his eyes that made her blood freeze.

Not grief for a lost colleague. Not even a concern for the research setback. There was satisfaction there, cold and calculating, as if Hans's death had not been a tragedy but a successful test.

"We're creating the future, Meine Tochter," he said softly, his voice carrying a note of reverence that chilled her to the bone. "And sometimes, the future requires sacrifice."

As the containment team moved Hans's body onto a gurney, Eva noticed something that would haunt her dreams for weeks to come. On the lab bench where the accident had occurred, a single drop of Sample 7-Alpha had splashed onto a potted geranium—a small touch of life that one of the researchers had brought in to brighten the sterile environment.

The plant was dead, its leaves blackened and withered, but as Eva watched in fascination and horror, new growth was already beginning to emerge. Dark, twisted stems that looked nothing like the original plant, reaching toward the fluorescent lights with an almost predatory hunger. The flowers that budded from these alien stalks were wrong—too symmetrical, too perfect, as if they'd been designed rather than grown.

And at the base of the plant, she could see something that made her stomach lurch. Tiny, hair-like tendrils were spreading outward from the root system, seeking new sources of organic matter to convert and transform.

Whatever they were creating here, it wasn't a cure for anything.

It was something else entirely. Something that consumed life and rebuilt it according to its own twisted design.

As she watched the security team finish their cleanup, Eva made a decision that would change everything. She'd discover what Sample 7-Alpha truly was, what her father's genuine intentions were, and what connection this project had to the encrypted files she'd glimpsed on his computer.

She would discover the truth about the Prometheus Project.

Even if it killed her.

Behind her, hidden in the shadows of the corridor, a security camera tracked her every movement. In a control room three floors below, Wilhelm von Hess watched the footage with interest, his fingers steepled before him.

"She's beginning to ask the right questions," he murmured to Klaus, who had joined him after the incident. "How much longer can we maintain the deception?"

Klaus's face was tight with worry. "She's brilliant, but she's also idealistic. Like her mother was. If she learns the truth..."

"Then we'll deal with her the same way we dealt with her mother," von Hess said calmly. "The Phoenix Order cannot afford sentiment, Klaus. Not even for family."

On the screen, Eva was walking back to her laboratory, her shoulders set with determination. She had no idea that every move was being monitored, her every doubt catalogued and analyzed.

The future that Klaus had promised was coming.

And Eva Richter would either embrace it or be consumed by it.

# Chapter 3: Fragments of Truth

Ben Carter's morning routine had become a ritual of paranoia disguised as academic normalcy. He woke at 5:47 AM—never setting an alarm, his internal clock honed by years of intelligence work—and immediately checked the micro-threads he'd placed across his apartment's entry points. Hair-thin fishing line, nearly invisible, positioned at precise angles that would reveal any intrusion during the night.

All intact. As they had been for three years.

The apartment's windows faced east, providing natural surveillance of the street below while he prepared his coffee. French press, single-origin beans from a small roaster in Kreuzberg—the kind of detail that made him seem like just another pretentious academic. But the real reason was tactical: the brewing process gave him exactly four minutes to scan the street for watchers, anomalies, patterns.

Today, something was different.

A white van had been parked across the street for two days now. Different license plates each morning, but the same dent in the rear bumper. The kind of detail most people would miss. The sort of detail that had kept him alive in Damascus, Belgrade, and a dozen other places where American intelligence officers weren't supposed to exist.

Ben's hands were steady as he poured the coffee, but his mind was racing. The encrypted message from yesterday hadn't been a coincidence. Sarah Chen was trying to warn him about something, and if Sarah was concerned enough to break three years of silence, then Ben's carefully constructed new life was about to crumble.

The walk to the university took him through Charlottenburg's tree-lined streets, past the same cafés and bookshops he'd frequented for three years. But today he varied his route, taking an indirect path that would reveal any surveillance. Old habits, muscle memory, the kind of tradecraft that had once been second nature.

His office at Friedrich-Wilhelm University was on the fourth floor of the History Department, a cramped space filled with books that served as both academic props and security measures. The positioning wasn't accidental—thick volumes of German military history provided sound dampening. At the same time, the strategic placement of mirrors allowed him to observe both the corridor and the courtyard below.

But as he climbed the stairs, Ben noticed something that made his blood run cold. The department secretary, Frau Kellner, was at her desk unusually early, her movements too deliberate, too careful. She looked up as he passed, offering a smile that didn't quite reach her eyes.

"Guten Morgen, Professor Carter. You have a visitor waiting in your office."

Ben's hand moved instinctively toward his jacket pocket, where a weapon should have been. "A visitor? I don't have any appointments scheduled."

"A colleague from the university," she said, her voice carrying a slight tremor. "She said it was urgent."

The door to his office was slightly ajar—never how he left it. Ben pushed it open carefully, his body positioned to retreat if necessary. Inside, a woman sat in his chair, her back to the door, dark hair pulled into a severe bun.

"Hello, Ben."

He recognized the voice immediately. Sarah Chen, his former handler, the woman who'd recruited him fresh out of graduate school and turned him into one of the CIA's most effective intelligence analysts. She turned slowly, and Ben saw that three years of official retirement had aged her more than the previous decade of intelligence work.

"Sarah." Ben stepped inside and closed the door, engaging the lock with a soft click. "I thought you were enjoying your pension."

"Retirement is overrated." She stood, her movements betraying the tension Ben recognized from their old operational days. "We need to talk."

"About what? I've been out of the game for three years. I teach history to bored undergraduates and grade papers about the Marshall Plan. Whatever this is about, I'm not interested."

Sarah's smile was grim. "I'm afraid it's not that simple. Tell me, Ben, have you been experiencing any unusual... inquiries lately? People asking about your background, your previous work?"

Ben's mind flashed to the strange van, Frau Kellner's nervous behavior, the encrypted message. "Maybe. Why?"

"Because your name has been appearing in some very disturbing intelligence chatter. Extremist networks, international terror cells, people who shouldn't know you exist." She moved to the window, peering through the blinds at the courtyard below. "Someone's been asking questions about Dr. Benjamin Carter, former CIA intelligence analyst, expert in Middle Eastern terror networks."

The past Ben had tried so hard to bury came flooding back. Damascus, 2021. The intelligence report that had seemed so clean, so actionable. A terror cell is planning an attack on a refugee camp. GPS coordinates, behavioral patterns, and communication intercepts. Everything had pointed to a legitimate target.

"Sarah, whatever this is about, I can't help you. I'm done with that life."

"Are you?" She turned from the window, her eyes hard. "Because that life isn't done with you. There's speculation about a group known as the Phoenix Order. Ring any bells?"

Ben's carefully constructed composure cracked slightly. "No. Should it?"

"Maybe. Maybe not. But they seem to know about you. About Damascus. About what happened there."

The nightmare came back unbidden. The drone strike he'd recommended, the target he'd verified. The wedding party that intelligence had missed, the thirty-seven civilians who'd died alongside the terror cell. Ben had been thorough, analytical, and careful. But not careful enough.

"The Syrian operation was classified at the highest levels," Ben said quietly. "No one outside the agency should know about it."

"That's what worries me." Sarah moved closer, her voice dropping. "Someone's been digging into your past, Ben. Deep digging. The kind that suggests either a massive intelligence leak or someone with extraordinary access to classified information."

Ben's phone buzzed. Another encrypted message, this one from a different number:

Your Damascus report created opportunities. We remember Syria. We remember you. —Phoenix

His blood went cold. Only a handful of people were aware of the specific details of the Syrian operation. The civilian casualties had been buried in classified reports, the entire incident sanitized from official records. If someone outside the intelligence community had access to that information...

"Sarah, who exactly is the Phoenix Order?"

"That's what we're trying to figure out. What we know is limited but disturbing. Extreme right-wing ideology, significant financial backing, and apparent access to classified intelligence. They've been recruiting scientists, former military personnel, individuals with specific skill sets."

Ben's mind began processing the information with the analytical precision that had made him valuable to the agency. "What kind of skill sets?"

"Bioweapons research. Genetic engineering. Viral pathology." Sarah's expression darkened. "And intelligence analysis. Specifically, analysts with expertise in target identification and operational planning."

The implications hit Ben like a physical blow. "They want to use my skills."

"Or they want to frame you for using them." Sarah pulled out a tablet, showing him a series of encrypted communications. "Look at this. These messages were intercepted from a secure Phoenix Order channel. Your name appears seventeen times in the past month alone."

Ben scanned the messages, his trained eye picking up the patterns. References to "the Asset," "Damascus protocols," and "civilian acceptability thresholds." Someone was using his analytical framework and methodologies to plan something monstrous.

"Sarah, I need to know everything. What is the Phoenix Order planning?"

"We don't know. That's why I'm here." She moved to the door, checking the corridor through the keyhole. "But there's something else. Intelligence chatter mentions a virus known as the Lethe Virus. Bio-weapons research, genetic targeting, mass casualty potential."

Ben's academic knowledge clicked into place. "Lethe. The river of forgetfulness in Greek mythology. Souls drank from it to forget their past lives before reincarnation."

"Appropriate name for a weapon designed to erase populations," Sarah said grimly. "Ben, I need you to come back. Officially, you're still on administrative leave. Unofficially, we need your analytical skills to—"

"No." Ben's voice was firm. "I'm done with that world. I'm done with being responsible for people's deaths."

"And if your inaction results in thousands more deaths?"

The question hung in the air between them. Ben thought of his students, their earnest faces, and the questions they had about moral courage. How could he lecture them about the importance of taking a stand while hiding from his own responsibilities?

His phone buzzed again. This time, the message was longer:

Dr. Carter, you don't know me, but my name is Eva Richter. I'm a biochemist working on a project that I believe is connected to the Phoenix Order. People are going to die unless we stop them. I need your help. I have information about the Lethe Virus. Meet me at the Café Einstein on Unter den Linden tomorrow at 2 PM. Come alone. If you don't come, millions will die. —E.R.

Ben stared at the message, his mind racing. Eva Richter. The name wasn't familiar, but the desperation in her words was genuine. And if she had information about the Lethe Virus...

"Sarah, I need to show you something."

Before he could display the message, Sarah's phone buzzed urgently. She glanced at it, and her face went pale.

"We have to go. Now."

"What's wrong?"

"Surveillance teams are moving on this location. Someone knows I'm here." She moved to the window, her trained eyes scanning the courtyard. "Ben, whoever the Phoenix Order is, they have resources and reach that go far beyond what we initially estimated."

Ben's paranoia crystallized into certainty. The white van, Frau Kellner's nervous behavior, the timing of this meeting was all orchestrated. Someone had wanted Sarah to find him had wanted this conversation to happen.

"It's a trap," he said quietly. "They're using you to flush me out."

"Maybe. Or maybe they're using me to get to you." Sarah's voice was tight with professional concern. "Ben, I need you to listen carefully. Whatever's happening here, it's bigger than both of us. The Phoenix Order isn't just another extremist group. They have government connections, unlimited funding, and access to classified intelligence."

The sound of footsteps in the corridor made them both freeze. Multiple sets, moving with tactical precision. Ben grabbed his go-bag from behind the bookshelf, his mind switching to operational mode.

"Fire escape," he whispered. "Through the window."

But as they moved toward the window, Ben caught sight of the courtyard below. Black SUVs, tactical teams, the kind of coordinated response that suggested government backing. This wasn't some rogue terrorist cell. This was a professional intelligence operation.

"Sarah, who exactly sent you to find me?"

Her silence was answer enough.

Ben's world tilted. His former handler, the woman who'd been like a mentor to him, had led his enemies directly to him. Whether intentionally or not, she'd compromised his location, his safety, his carefully constructed new life.

"I'm sorry, Ben," she said quietly. "I thought I could protect you."

The door to his office exploded inward. Tactical teams poured through, weapons drawn, shouting commands in German and English. Ben had a split second to make a decision that would determine whether Dr. Benjamin Carter, the professor, would survive the next ten minutes.

He chose survival.

The window glass shattered as Ben dove through it, his body rolling across the fire escape platform. Behind him, he could hear gunfire, Sarah's voice shouting, the chaos of a compromised intelligence operation.

As he rappelled down the fire escape, Ben's mind was already working through the implications. The Phoenix Order knew about Damascus. They had government connections. They were using his analytical methods to plan something involving biological weapons.

And now they were hunting him.

At the bottom of the fire escape, Ben looked up at his office window. Sarah's face appeared briefly, blood streaming from a head wound, her eyes meeting his one last time. Then she was gone, pulled back into the darkness by unseen hands.

Ben ran. Through the university campus, past the students who had no idea their professor was fleeing for his life, into the Berlin streets where he would have to become someone else entirely.

The Phoenix Order had found him.

Now he had to find Eva Richter before they found her too.

The encrypted message mentioned millions of deaths. Ben had thought he was done with preventing catastrophes.

But catastrophes, it seemed, weren't done with him.

# Chapter 4: The Watcher

E va Richter couldn't sleep. For the third night in a row, she lay in her small quarters within the monastery complex, staring at the ceiling while images of Hans Mueller's transformation played on endless repeat behind her closed eyes. The way his skin had changed, darkened, and became something alien. The inhuman sounds he'd made. The predatory intelligence that had flickered in his eyes just before the end.

And that plant. The twisted, wrong growth that had sprouted from the contaminated geranium, reaching toward the light with an almost conscious hunger.

At 3:17 AM, she gave up on sleep and padded barefoot to her small kitchenette. The monastery's ancient stones retained the warmth of the day, but Eva felt chilled to the bone. She made tea—chamomile with honey, the same combination her mother had prepared during Eva's childhood nightmares—and settled at her desk with Hans's personnel file.

Hans Mueller, age twenty-six, PhD in biochemistry from ETH Zurich, specialized in viral pathology. Recruited six months ago, at the same time as three other researchers who'd all been assigned to different projects within the facility. All young, all brilliant, all idealistic. All are working on aspects of the same research without knowing it.

Eva's training had taught her to think systematically, to identify patterns, and to question inconsistencies. And the more she examined the structure of their research, the more inconsistencies she found.

The Helix Dynamics facility housed fourteen different research teams, each working on projects that appeared to be separate. But Eva's access to the central databases had allowed her to glimpse the bigger picture. The protein structures she'd been optimizing, the cellular delivery mechanisms, and the genetic targeting systems were all components of a single, massive undertaking.

She opened her laptop and began tracing the connections. Team Seven was working on rapid cellular replication. Team Nine was developing genetic markers for population-specific targeting. Team Twelve was creating airborne dispersal systems. Each team believed they were working on therapeutic applications, but taken together...

Eva's blood ran cold. Taken together, they were developing a bioweapon of unprecedented sophistication.

She needed to see the restricted sections of the facility. The areas her keycard couldn't access were the laboratories that Klaus had told her were focused on "theoretical research," which didn't concern her immediate work.

The monastery's corridors were dimly lit at night, emergency lighting casting long shadows between the ancient stone arches. Eva had grown up in these halls, knew every hidden passage and forgotten storeroom from her childhood visits to her father's work. But tonight, the familiar spaces felt alien, threatening.

Her keycard worked on the main laboratory doors. Still, when she tried to access Sub-Level 3, the reader flashed red and emitted a soft beep that seemed to echo through the silence. Access denied.

That was new. Six months ago, her clearance had been universal within the facility. Now, apparently, there were areas her father didn't trust her to see.

But Eva had advantages that her father had forgotten. As a child, she'd explored every inch of the monastery, including the medieval tunnels that connected the different wings. Maintenance passages that bypassed the modern security systems.

She found the entrance behind a false wall in the wine cellar. This narrow stone passage led to the facility's lower levels. The air was thick with the smell of ancient mortar and modern chemicals. This unsettling combination made her stomach churn.

Sub-Level 3 was a different world. The medieval stone gave way to brushed steel and reinforced glass. The lighting was harsh, clinical, and designed for precision work rather than comfort. And the security was of a military-grade standard, featuring motion sensors, pressure plates, and biometric scanners on every door.

Through the reinforced windows, Eva could see laboratories that looked more like weapons testing facilities than medical research stations. Massive containment units, automated systems for handling dangerous materials, and equipment she didn't recognize but which radiated an aura of purpose and menace.

In one laboratory, she saw something that made her heart race. A production line for synthesizing biological agents, complete with quality control stations and packaging systems. The scale was industrial, designed for mass production rather than research.

Whatever they were creating here, they intended to produce a large quantity of it.

Eva's hands shook as she took photographs with her phone, documenting everything she could see. The evidence was overwhelming, but she still didn't understand the full scope of what her father was involved in.

She needed to find his personal files, the research notes he kept separate from the official databases. She needed to understand how her brilliant, caring father had become part of something so monstrous.

The answer lay in his private office, a space she'd avoided since childhood. Klaus had always been protective of his personal workspace, claiming it contained sensitive material that wasn't appropriate for children. Now, Eva wondered what other secrets it might contain.

The office was on the monastery's third floor, accessible through Klaus's private elevator. Eva's keycard shouldn't have worked, but it did—another sign that her father still trusted her, despite the growing distance between them.

The office was exactly as she remembered from her rare childhood visits. Dark wood paneling, leather-bound books, and a massive oak desk that had belonged to her grandfather. But now Eva saw details that her child's mind had missed—the sophisticated computer systems, the encrypted communication devices, the safe built into the wall behind a false bookshelf.

She started with Klaus's computer, but it was password-protected with biometric locks she couldn't bypass. The desk drawers were locked as well, but Eva remembered her father's habits. He'd always been absent-minded about small security measures, relying on the facility's overall systems rather than personal vigilance.

She found the key taped to the underside of the center drawer of the desk, exactly where Klaus had hidden spare keys during her childhood.

The drawers contained the usual administrative materials—budgets, personnel files, and correspondence with funding agencies. But in the bottom drawer, Eva found something that made her breath catch.

A thick folder labeled "Ingrid Richter - Personal Research." Her mother's name, in her father's handwriting.

Eva opened the folder with trembling hands. Inside were pages of her mother's research notes, written in the careful script Eva remembered from childhood. But these weren't the cancer research papers Eva had seen before. These were something else entirely.

Her mother had been working on antiviral compounds, specifically targeting synthetic pathogens. The notes described defensive measures against engineered bioweapons, as well as methods for creating broad-spectrum countermeasures that could neutralize multiple viral strains simultaneously.

And scattered throughout the notes were references to something called "the Phoenix Protocol" and mentions of an intelligence analyst named Benjamin Carter.

Eva's hands shook as she read her mother's words:

"The Phoenix Order's goals are clear from Klaus's private communications. They intend to use the viral compound to eliminate what they consider 'genetically inferior' populations. The targeting mechanism is sophisticated enough to focus on specific ethnic markers while leaving others untouched. This is not medicine—it is genocide disguised as evolution."

The next entry was dated three weeks before her mother's death:

"I've made contact with CIA analyst Benjamin Carter. His expertise in population targeting and strategic analysis could be crucial for understanding the Phoenix Order's operational plans. Carter has agreed to help develop countermeasures, but we must be careful. Klaus suspects my true loyalties."

The final entry was written in her mother's increasingly shaky handwriting:

"Klaus knows about my research. He's demanded that I destroy all my work on the antiviral compounds. I've hidden the critical formulas in places he'll never think to look. If something happens to me, Eva must understand that the counter-virus is possible. The key is in the lullaby I used to sing to her, the one about the phoenix rising from ashes. The genetic sequence is encoded in the melody—she'll know how to read it when the time comes."

Eva's world tilted. Her mother hadn't died of cancer. She'd been murdered. Killed by her own husband for trying to stop a genocide.

And the counter-virus, the weapon that could stop the Phoenix Order's plans, was hidden in her childhood memories.

The lullaby. Eva could hear it now, her mother's voice singing the ancient melody about rebirth and renewal. She'd thought it was just a song, a comforting routine from her childhood. But it was actually a genetic code, a formula for salvation encoded in music.

A sound in the corridor made Eva freeze. Footsteps, approaching the office. She quickly photographed the relevant pages and returned the folder to its hiding place, but she was too late.

The office door opened, and Klaus entered, flanked by two men in expensive suits. Eva recognized one of them from photographs—Wilhelm von Hess, the project's mysterious benefactor.

"Eva," Klaus said, his voice carrying disappointment rather than surprise. "I thought I might find you here."

"Papa, I can explain—"

"There's no need for explanations, Meine Tochter." Klaus moved to his desk, his movements careful and deliberate. "I know why you're here. You have questions about the project, about your mother's research, about what we're truly accomplishing."

Wilhelm von Hess stepped forward, his presence filling the room with an aura of casual menace. He was younger than Eva had expected, perhaps fifty, with the kind of aristocratic bearing that suggested old money and older power. His eyes were pale blue, almost colorless, and when he looked at Eva, she felt like a specimen under a microscope.

"Dr. Richter," von Hess said, his voice cultured, almost gentle. "You've been asking the right questions. That's admirable in a scientist, but potentially dangerous in our current circumstances."

"What circumstances?" Eva forced herself to stand straighter, to project a confidence she didn't feel. "What exactly is the Phoenix Order?"

Von Hess smiled, and Eva realized she'd just confirmed that she knew far more than she should. "The Phoenix Order is humanity's future, Dr. Richter. We are the architects of a new world, one free from the genetic corruption that has weakened our species for generations."

"Genetic corruption?" Eva's voice was steady despite her racing heart. "You're talking about ethnic cleansing."

"I'm talking about evolution," von Hess replied. "Natural selection is guided by scientific precision rather than random chance. The Lethe Virus will eliminate the weak, the inferior, the genetically compromised. What remains will be stronger, purer, more capable of surviving in the world we're creating."

Klaus moved to Eva's side, his hand resting on her shoulder in a gesture that had once been comforting but now felt like a threat. "Your mother never understood the necessity of what we're doing. She was brilliant, but she lacked the vision to see beyond immediate compassion to long-term benefit."

"You killed her," Eva said quietly. "She was trying to stop you, and you killed her."

"I saved her from making a terrible mistake," Klaus replied. "Just as I'm trying to save you now. Eva, you have a choice. You can join us willingly and use your talents to help create a better world. Or..."

He didn't finish the sentence. He didn't need to.

Von Hess moved closer, studying Eva with the intensity of a predator evaluating prey. "You have your mother's intelligence, but I hope you also have more practical wisdom. The Phoenix Order offers you the chance to be part of something greater than yourself. To help guide humanity's evolution rather than simply documenting its decline."

"And if I refuse?"

"Then you'll share your mother's fate," von Hess said matter-of-factly. "But I don't think you'll refuse. You're too intelligent to throw your life away on misguided idealism."

Eva looked at her father, searching his face for any sign of the man who'd raised her, who'd comforted her nightmares, who'd taught her to love science and discovery. But all she saw was cold calculation and absolute certainty.

"I need time to think," she said finally.

"Of course," von Hess replied. "But not too much time. The convergence is approaching, and we'll need your complete commitment very soon."

As Eva left the office, she could feel their eyes on her back, watching, evaluating, planning. She'd bought herself time, but at the cost of revealing how much she knew.

In her quarters, she sat at her desk and began to hum the lullaby her mother had sung. The melody was complex, with intricate variations that had seemed random in childhood but now revealed themselves as deliberate, structured, purposeful.

Her mother had been brilliant. She'd hidden the formula for the counter-virus in plain sight, encoded in music that only Eva would remember and understand.

But decoding it would require resources Eva didn't have, expertise she lacked, and allies she couldn't trust.

Except for one person. Benjamin Carter, the CIA analyst her mother had mentioned. If he were still alive and willing to help, he might be the key to stopping the Phoenix Order.

Eva opened her laptop and began composing an encrypted message. She had no idea if Carter was still at the contact information her mother had left, but she had to try.

The future of humanity might depend on it.

And outside her window, hidden in the shadows of the monastery's ancient walls, surveillance cameras tracked her every movement, recording her every expression for analysis by the Phoenix Order's psychological profilers.

The watcher was being watched.

And her time was running out.

# Chapter 5: The Warning

T he Café Einstein on Unter den Linden was a monument to old-world Berlin elegance, its wood-paneled walls and red velvet banquettes unchanged since the 1920s. Ben Carter had chosen it for the meeting because it offered multiple exit routes and good sight lines. Still, as he sat nursing his third cup of coffee, he couldn't shake the feeling that he was walking into a trap.

Sarah Chen was twenty minutes late. In their previous life, Sarah had been pathologically punctual, a trait that had served her well in the intelligence community, where timing could mean the difference between life and death. Her tardiness now suggested either extreme caution or extreme danger.

Probably both.

Ben had positioned himself at a corner table, with his back to the wall, offering a clear view of the entrance and quick access to the service corridor that led to the kitchen. Old habits, but habits that had kept him alive in Damascus, Belgrade, and a dozen other places where American intelligence officers weren't supposed to exist.

The café's lunchtime crowd provided good cover—businessmen conducting meetings, tourists consulting guidebooks, university students typing frantically on laptops. Normal people, very much in their everyday lives, oblivious to the fact that they were sharing space with two former intelligence operatives whose past was about to catch up with them.

At 2:23 PM, Sarah finally appeared.

Ben almost didn't recognize her. The confident, impeccably dressed handler he remembered had been replaced by a woman who looked like she'd aged a decade in three years. Her dark hair was streaked with premature gray, her clothes were wrinkled, and her eyes bore the haunted look of someone who had seen too much and slept too little.

She scanned the café with the systematic thoroughness of a trained operative, cataloging faces, exits, and potential threats. When her gaze found Ben, she didn't smile or wave. Instead, she gave an almost imperceptible nod and moved to the bar, ordering coffee while continuing her surveillance.

Five minutes later, she joined him at his table, sliding into the seat across from him with her back to the room—a violation of basic tradecraft that told Ben everything he needed to know about her mental state.

"You look terrible," Ben said quietly.

"Retirement doesn't suit me." Sarah's voice was hoarse, as if she'd been shouting or hadn't spoken to another human being in days. "Thanks for coming."

"I almost didn't. After yesterday's excitement at the university, I figured my academic career was over anyway."

Sarah's eyes sharpened. "What happened yesterday?"

"You don't know?" Ben studied her face, looking for deception. "Tactical teams raided my office. I barely escaped through the fire escape. I assumed you were somehow involved."

"I wasn't." Sarah's denial was immediate and convincing. "Ben, I've been completely off-grid for the past seventy-two hours. No official contact, no agency communication, nothing. Whatever happened at the university, it wasn't sanctioned by anyone I know."

The implications were chilling. If Sarah wasn't behind the raid, then someone else had tracked Ben to the university. Someone with government resources and tactical capabilities.

"Sarah, what exactly is going on?"

She leaned forward, her voice dropping to barely above a whisper. "Ben, you can't hide from the world forever. Evil doesn't pause for your conscience—it grows stronger while good people do nothing. Sooner or later, you have to choose a side."

"I chose my side three years ago when I walked away from the agency. I chose peace, teaching, a life where my analysis doesn't get people killed."

"And how's that working out for you?" Sarah's smile was bitter. "Your apartment's been under surveillance for weeks. Your phone's been tapped. Your university colleagues have been questioned by people claiming to be federal auditors. You think you've been living a peaceful life, but you've actually been living in a cage."

Ben's coffee cup rattled against the saucer as he set it down. "Surveillance by whom?"

"That's what I'm trying to figure out." Sarah glanced around the café, her paranoia infectious. "But I've got fragments, pieces of intelligence that don't add up to a complete picture. The Phoenix Order, the Lethe Virus, your name appearing in communications that shouldn't exist."

"Tell me about the Phoenix Order."

Sarah opened her mouth to respond, then stopped. Her eyes had fixed on something over Ben's shoulder, and her face went pale.

"We need to leave. Now."

"Sarah, what—"

"Now, Ben. Move."

The urgency in her voice triggered Ben's operational instincts. He was already standing, moving toward the service corridor, when the first shot shattered the café's front window.

The bullet took Sarah in the shoulder, spinning her around and sending her crashing into the neighboring table. Chaos erupted instantly, with screaming customers, overturned chairs, and the bitter smell of gunpowder mixing with the café's rich coffee aroma.

Ben dove behind the bar as the second shot shattered the mirror behind him. Through the broken window, he could see the muzzle flash from a building across the street—a professional sniper, perfectly positioned to cover the café's main entrance.

"Sarah!" Ben called out, but there was no response. He could see her crumpled form near the overturned table, blood pooling beneath her still body.

The sniper had been waiting for them. This wasn't a random attack or a robbery gone wrong. This was a professional assassination, timed to eliminate Sarah before she could reveal whatever she'd come to tell him.

Ben's mind raced through his options. The front entrance was covered by the sniper. The kitchen exit might be compromised. The bathroom window was too small for an adult to fit through. But the service corridor led to a basement storage area that connected to the building's underground parking garage.

He made his decision.

Moving in a crouch, Ben worked his way through the panicked crowd toward the service corridor. The café's patrons were screaming, diving for cover, calling for help on their phones. Perfect chaos to cover his escape.

Another shot rang out, and Ben realized the sniper was systematically eliminating witnesses. This wasn't just about Sarah—it was about preventing any information from reaching him.

The service corridor was narrow and dimly lit, lined with supplies and equipment. Ben moved quickly but carefully, his senses heightened to combat-level awareness. Every shadow could hide a threat; every sound could indicate an ambush.

At the end of the corridor, he found the basement access door. It was locked, but Ben's training had included lock-picking among other less savory skills. Thirty seconds later, he was through the door and into the building's basement.

The basement was a maze of storage areas, utility rooms, and maintenance passages. Ben oriented himself by memory and instinct, working his way toward what he hoped was the parking garage. Behind him, he could hear footsteps—heavy boots, multiple sets, moving with tactical precision.

They'd found his route.

Ben picked up his pace, moving as quickly as he dared through the unfamiliar basement. The footsteps behind him were getting closer, and he could hear radio chatter—professional communications, coordinated pursuit.

He reached the garage access door just as the first tactical team rounded the corner behind him. The door was made of heavy steel, reinforced for security, and was also electronically locked.

Ben's training had included basic electronics, but not enough to bypass a sophisticated lock system. He was trapped.

Then he heard the voices behind him—not German, not English, but a language he recognized from his intelligence work. Russian. The team hunting him was speaking Russian.

That changed everything. This wasn't a German police response or even a CIA operation. This was something entirely different, suggesting international connections and resources that went far beyond what he'd imagined.

The garage door's electronic lock disengaged with a soft click. Ben spun around, expecting to see his pursuers, but instead found himself face-to-face with a young woman in a laboratory coat.

"Dr. Carter?" Her voice was accented German, but with the precise diction of someone speaking a second language. "My name is Eva Richter. I sent you the encrypted message."

Ben stared at her in shock. "How did you—"

"Find you? I've been tracking your movements since yesterday. I know about the university raid, about your former handler, about the Phoenix Order's interest in your background." Eva's eyes were bright with intelligence and a hint of desperation. "We need to talk, but not here. They're closing in on both of us."

The sound of approaching footsteps made them both freeze. Eva moved to the garage door's control panel, her fingers dancing across the keypad with practiced efficiency.

"I have a car," she said. "But we need to move now."

Ben hesitated. Everything about this was wrong: the timing, the location, the coincidence of Eva's appearance. But the alternative was capture by the Russian team, and his instincts told him that would be considerably worse than trusting a mysterious German scientist.

"Where?" he asked.

"Safe house. I'll explain everything, but we need to get out of Berlin immediately."

The garage door slid open, revealing a BMW sedan with its engine running. Eva moved toward it with the confidence of someone who'd planned this extraction carefully.

"Ben," she said, turning back to him. "I know this is difficult to believe, but I'm probably the only person in the world who can help you stop what's coming. The Phoenix Order isn't just interested in your analytical skills—they're using your Damascus protocols to plan something that will make Syria look like a minor incident."

The footsteps were getting closer. Ben could hear Russian voices, urgent and coordinated. In seconds, the tactical team would reach the garage.

"What's coming?" Ben asked.

"Global genocide," Eva replied. "Disguised as a pandemic. And we have maybe seventy-two hours to stop it."

Ben made his decision. He followed Eva to the car, and as they drove out of the garage, he caught a glimpse of the tactical team emerging from the basement—black uniforms, professional equipment, faces hidden behind tactical masks.

In the passenger seat, Eva handed him a tablet showing satellite imagery of a monastery complex in the Bavarian Alps.

"That's where they're planning to launch the attack," she said. "That's where my father has been developing the Lethe Virus. And that's where we need to go if we want to save approximately four billion people."

Ben stared at the imagery, his analytical mind processing the implications. "Your father?"

"Dr. Klaus Richter. Brilliant scientist, loving father, and apparently a genocidal monster." Eva's voice was steady, but Ben could hear the pain beneath the surface. "He killed my mother for trying to stop this. Now he's planning to kill most of the human race."

As they drove through Berlin's streets, Ben realized that his quiet academic life was truly over. The Phoenix Order had found him; his former handler was dead, and he was now partnered with a woman whose family connections would either be his salvation or his doom.

"Eva," he said quietly. "Why should I trust you?"

"Because," she replied, "I'm the only one who knows how to stop the virus. My mother hid the formula for the counter-agent before she died. It's encoded in a lullaby she used to sing to me."

Ben looked at her profile as she drove, seeing the intelligence in her eyes, the determination in her jawline, the weight of responsibility on her shoulders. She was brilliant, desperate, and possibly their only hope.

"A lullaby?"

"My mother was very clever. She knew they'd search for hidden files, encrypted data, and secret laboratories. But they'd never think to look for a genetic formula disguised as a children's song."

As they left Berlin behind, heading toward the Alps. Whatever lay ahead, Ben realized that his academic expertise in recognizing evil was about to be put to the test in the most practical way possible.

The Phoenix Order was rising.

And he was finally ready to choose a side.

# Chapter 6: The Kill Switch Discovery

Eva Richter sat in her father's office at 3:47 AM, the monastery's ancient stones holding the Alpine night's chill despite the modern heating system. Sleep had become impossible since Hans Mueller's death three days ago. Every time she closed her eyes, she saw his transformation, heard his inhuman screams, witnessed the predatory intelligence that had flickered in his gaze before the end.

But it wasn't just the horror of Hans's death that kept her awake. It was the questions that multiplied like viruses, each answer spawning three new uncertainties. Why had her keycard access been mysteriously restricted? Why did the facility's security protocols seem designed for warfare rather than medical research? And why did her father's eyes hold such cold satisfaction when he spoke of Hans's "successful test"?

The answers lay behind the encrypted files she'd glimpsed on Klaus's computer, files that had been haunting her thoughts for weeks. Tonight, she would finally discover the truth.

Klaus's office was exactly as she remembered from childhood visits—dark wood paneling, leather-bound books, the massive oak desk that had belonged to her grandfather. But now Eva saw details that her child's mind had missed: the multiple computer screens, the encrypted communication devices, the subtle surveillance equipment that suggested this wasn't just a scientist's workspace but a command center.

The computer was password-protected, but Eva had advantages Klaus had forgotten. She'd been his daughter for twenty-eight years, had watched him work, had learned his habits and patterns. The man who could develop the world's most sophisticated bioweapon was still absent-minded about personal security.

She tried the obvious passwords first: his birthdate, her mother's name, significant dates in his career. Nothing. The security system allowed three attempts before triggering lockdown protocols, and Eva had used two.

Think, she told herself. What would Klaus use that was personal enough to remember but private enough to protect?

Then it came to her. Her mother's maiden name combined with the date of Eva's first published paper—the achievement that had made Klaus proudest, the moment when he'd realized his daughter had inherited his scientific brilliance.

The screen unlocked with a soft chime.

Eva's hands trembled as she navigated through Klaus's files. The directory structure was complex, layered with multiple security protocols, but she found what she was looking for in a folder labeled "Prometheus - Phase VII."

She opened the first file and felt her world tilt.

LETHE VIRUS - OPERATIONAL PARAMETERSClassification: PROJECT PHOENIX - EYES ONLYEstimated Casualties: 4.2 billion (preliminary targeting)Target Demographics: Sub-Saharan Africa (94% mortality), South Asia (87% mortality), Latin America (82% mortality)Genetic Markers: [DETAILED TECHNICAL SPECIFICATIONS FOLLOW]

Eva's breath caught. This wasn't medical research. This was genocide disguised as science.

She scrolled through the technical specifications, her trained eye recognizing the elegant horror of what her father had created. The virus targeted specific genetic markers, DNA sequences that correlated with geographic ancestry and ethnic heritage. It was designed to kill billions while leaving others virtually untouched.

The delivery system was equally sophisticated. Aerosol dispersal through major airport ventilation systems, timed to coincide with peak holiday travel. The virus would be carried worldwide by unknowing passengers, spreading to their families, their communities, their nations.

Eva found the target list and felt bile rise in her throat:

SIMULTANEOUS RELEASE POINTS:

- John F. Kennedy International Airport, New York

- Heathrow Airport, London

- Charles de Gaulle Airport, Paris

- Frankfurt Airport, Germany

- Dubai International Airport, UAE

- Tokyo Haneda Airport, Japan

- Los Angeles International Airport, USA

- São Paulo International Airport, Brazil

- OR Tambo International Airport, South Africa

- Mumbai International Airport, India

- Beijing Capital International Airport, China

- Sydney Kingsford Smith Airport, Australia

RELEASE DATE: December 24, 2024 - 1200 GMT

Christmas Eve. Four days away.

Eva's hands shook as she continued reading. The operational plan was breath-takingly comprehensive, accounting for every variable, every contingency. Airport security protocols, passenger flow patterns, ventilation system specifications, and even weather conditions that might affect dispersal efficiency.

But it was the following file that made her blood freeze.

OPERATION DAMASCUS - ANALYTICAL FRAMEWORKPrepared by: Dr. Benjamin Carter, CIA Intelligence Analysis DivisionAdapted for Phoenix Operations by: Dr. Klaus Richter

Eva opened the file and found herself looking at what appeared to be a legitimate intelligence report. Population density analysis, target selection criteria, and acceptable casualty thresholds. But the subject wasn't a Syrian terrorist cell—it was global population centers, billions of human beings reduced to statistical data points.

Ben Carter's analytical methods had been perverted, his frameworks for ethical intelligence operations twisted into tools for genocide. Eva could see the elegance of his original work, the careful consideration of civilian casualties, and the moral constraints that had guided his analysis. But Klaus had stripped away the ethical considerations, leaving only the cold mathematics of mass murder.

Another file caught her attention: ASSET RECRUITMENT - B. CARTER.

Eva opened it and found a psychological profile that made her stomach churn:

The subject exhibits classic symptoms of operational guilt following the Damascus incident. Thirty-seven civilian casualties have rendered him psychologically vulnerable to manipulation. Recommend an approach through academic channels, with gradual exposure to the Phoenix ideology, and eventual recruitment as a willing participant. If recruitment fails, the subject can be effectively framed as the architect of the operation, using his own analytical methods.

Assessment: Dr. Carter will either join us willingly or be destroyed by his own conscience. Either outcome serves Phoenix's interests.

They were planning to use Ben Carter as a scapegoat. If he didn't join them, they would use his own work against him to bring him down.

Eva continued reading, each file more horrifying than the last. The Phoenix Order wasn't just a terrorist organization—it was a global conspiracy with government connections, unlimited funding, and a carefully orchestrated plan to reshape human civilization.

But it was the personal files that broke her heart.

SUBJECT: Eva Richter - Operational Assessment

Eva's hands trembled as she read her father's clinical evaluation of her own psychology:

Eva exhibits typical idealistic patterns consistent with her mother's psychological profile. Strong moral compass, emotional attachment to humanitarian goals, vulnerable to manipulation through family loyalty. Her scientific brilliance makes her invaluable to the project, but her emotional instability requires careful management.

Recommend continued compartmentalization of information until the operational phase. Eva's work on protein synthesis is critical for viral efficiency. Still, she must remain unaware of the weapon's true purpose until deployment is irreversible.

Note: Eva's genetic profile indicates a 73% probability of survival following release. She will be useful in the post-convergence world, assuming proper ideological alignment can be achieved.

Eva stared at the screen in horror. Her father had been manipulating her for years, using her love for him and her mother's memory to turn her into an unwitting accomplice to genocide. Every conversation, every shared memory, every moment of paternal affection had been calculated to maintain her cooperation.

But the final file was the most devastating:

INGRID RICHTER - TERMINATION REPORT

Eva's mother hadn't died of cancer. She'd been murdered.

The report was clinical, detached, describing Ingrid's growing suspicions about the project and her attempts to contact outside authorities. Eva read about her mother's desperate efforts to develop countermeasures, her secret communications with intelligence analysts, her growing realization that her husband was planning humanity's destruction.

The subject became increasingly unstable following the discovery of Phase VI parameters. Attempted to contact CIA analyst Benjamin Carter with classified information about Phoenix operations. Termination authorized by Wilhelm von Hess. Death attributed to aggressive cervical cancer, consistent with the subject's medical history.

Note: Subject's research on viral countermeasures has been secured. Counter-virus development capabilities eliminated.

Eva's world collapsed. Her brilliant, loving mother had been murdered for trying to save humanity. The cancer that had destroyed their family, the slow, agonizing death that had driven Eva into biochemistry—it had all been a lie.

But Ingrid had been fighting back. The report mentioned hidden research, countermeasures that Klaus thought he'd destroyed. Eva remembered her mother's final months, the way she'd seemed to be working on something secret, something that gave her hope even as the "cancer" consumed her.

The lullaby. The childhood song about the phoenix rising from ashes, the melody that had seemed so random, so meaningless. Her mother had been trying to tell her something, to hide something in plain sight.

Eva was so absorbed in the files that she almost missed the soft chime from the computer's security system. An alert flashed on the screen: UNAUTHORIZED ACCESS DETECTED - SECURITY PROTOCOLS INITIATED.

Her blood went cold. The system had been monitoring her access, recording everything she'd viewed. Klaus would know within minutes that she'd discovered the truth.

Eva quickly copied the critical files to an encrypted drive, her hands shaking as she worked. She had maybe five minutes before security teams arrived, maybe less.

As she prepared to leave, one final file caught her attention: PHOENIX PER-SONNEL - ACTIVE ASSETS.

She opened it and found a list of names that made her knees weak. Government officials, military commanders, intelligence operatives, scientists, journalists—hundreds of people in positions of power and influence, all working for the Phoenix Order.

The conspiracy was vast, sophisticated, and nearly unstoppable. But at the bottom of the list, she found something.

# Chapter 7: The Escape

The file transfer progress bar crawled across the screen, indicating 73% completion. Eva's hands shook as she watched the encrypted data being copied to her portable drive, each percentage point representing seconds she didn't have. The security system's soft chimes had become more frequent, and she could hear the distant sound of elevator doors opening in the monastery's main corridor.

They were coming.

78% complete.

Eva's mind raced through the facility's layout, calculating escape routes she'd memorized as a child. The maintenance tunnels behind the wine cellar, the hidden passages her mother had shown her during those long-ago visits when Klaus had seemed like a loving father rather than a genocidal monster.

82% complete.

The sound of footsteps echoed through the corridor outside Klaus's office—multiple sets, moving with military precision. Eva recognized the rhythm from her childhood: security teams conducting facility sweeps, their movements coordinated and efficient. But this time, they were hunting her.

87% complete.

Eva's phone buzzed with an incoming message. She glanced at the screen and felt her blood freeze:

Eva, we need to talk. Office. Now. Don't make this harder than it needs to be.
—Papa

He knew. Klaus knew exactly what she'd discovered, exactly where she was, exactly what she was doing. The security system hadn't just been monitoring her access—it had been transmitting everything in real-time to his personal device.

91% complete.

The footsteps stopped outside the office door. Eva could hear voices—Klaus's familiar baritone mixed with others she didn't recognize. Official voices. Government voices. The kind of voices that made people disappear.

"I know she's in there," Klaus said, his tone carrying the patient authority of a man accustomed to being obeyed. "Give me five minutes to talk to her before we proceed with containment protocols."

Containment protocols. Eva's stomach churned. Hans Mueller had been subject to containment protocols.

96% complete.

The door handle began to turn. Eva grabbed the drive, pocketed it, and dove behind Klaus's massive oak desk just as the office door opened. Through the narrow gap between the desk and the wall, she could see her father's legs as he entered, followed by four other pairs of feet in military-style boots.

"Eva?" Klaus's voice was gentle, paternal, tinged with the kind of disappointment that had made her feel guilty as a child. "I know you're here, Meine Tochter. The security system tracks all movement within the facility."

Eva pressed herself against the wall, her heart hammering so loudly she was certain they could hear it. The desk provided temporary concealment, but Klaus knew his office better than anyone. He would find her within minutes.

"You've been reading my files," Klaus continued, his footsteps moving slowly around the office. "I understand your shock, your confusion. The scope of what we're accomplishing can be overwhelming, even for someone with your scientific background."

One of the other voices spoke—accented English, with a vaguely Eastern European tone. "Dr. Richter, we don't have time for family therapy. The security breach has compromised operational security. We need to proceed with immediate containment."

"She's my daughter," Klaus replied, his voice carrying a warning. "She'll be given the same choice we offered her mother."

Eva's breath caught. The same choice. Join willingly or be eliminated.

"Eva," Klaus said, his voice closer now. "I know you're frightened. I know what you've read seems monstrous. But you're seeing only fragments of a much larger picture. The Phoenix Order isn't about hate—it's about evolution. About guiding humanity toward a better future."

Eva could see Klaus's feet approaching the desk. In seconds, he would discover her hiding place. She had to move.

The office's layout flashed through her memory—the hidden panel behind the bookshelf. This passage led to the monastery's medieval tunnels. Her mother had shown her the entrance during one of their childhood games, back when the monastery had seemed like a magical place rather than a fortress of evil.

"The current world is dying, Eva," Klaus continued, his voice taking on the fervent tone she'd heard in his late-night conversations with Wilhelm von Hess. "Overpopulation, genetic degradation, the inevitable collapse of civilized society. The Lethe Virus isn't a weapon—it's a surgical tool, designed to remove the diseased tissue so the healthy organism can survive."

Eva spotted her chance. Klaus had moved to the window, his back to the desk. She slipped out from behind the oak furniture and toward the bookshelf, her movements silent despite her racing heart.

"The deaths will be swift, painless," Klaus said, his reflection visible in the dark window. "Far more merciful than the slow starvation and social collapse that would occur otherwise. We're not monsters, Eva. We're surgeons, preparing to save the patient."

Eva reached the bookshelf and found the hidden mechanism her mother had shown her—a seemingly decorative brass fitting that activated the passage when pressed in a specific sequence. The panel slid open with a soft whisper, revealing the narrow tunnel beyond.

"Your mother never understood," Klaus continued, his voice becoming harder. "She was brilliant, but she lacked the vision to see beyond immediate compassion to long-term necessity. I had hoped you would be different."

Eva was halfway through the passage when Klaus turned around.

"Eva!" His voice exploded across the office, no longer paternal but commanding. "Stop! You don't understand what you're doing!"

She ran. The tunnel was narrow, carved from medieval stone, lit only by emergency lighting that cast eerie shadows on the walls. Behind her, she could hear Klaus shouting orders, the sound of furniture being overturned, and the heavy boots of security teams entering the passage.

The tunnel split into multiple directions—a maze of medieval passages that connected different parts of the monastery. Eva chose the path that led toward the lower levels, toward the exit she'd used as a child when she'd wanted to explore the Alpine forest without her father's knowledge.

However, the security teams were also familiar with the passages. She could hear them behind her, their coordination professional and efficient. They were herding her, using the tunnel system to limit her escape routes.

Eva reached a junction and paused, listening. Footsteps approached from two directions, but the third passage—the one that led to the old wine cellar—seemed clear. She took it, moving as quickly as she dared in the dim light.

The wine cellar was ancient, its stone walls lined with empty racks that had once held the monastery's collection. But Eva wasn't interested in the wine storage. She was looking for the drainage tunnel that led to the outside world.

She found it behind a false wall, exactly where her mother had shown her twenty years ago. The tunnel was narrow, barely wide enough for an adult, but it led to a concealed exit in the forest beyond the monastery's walls.

Eva was crawling through the drainage tunnel when she heard Klaus's voice echoing through the wine cellar, closer than she'd expected.

"Eva, please! You're forcing me to make a choice I don't want to make!"

She could hear him clearly, which meant he was standing directly above her hiding place. The false wall wasn't as concealed as she'd hoped.

"I know you're in there," Klaus continued, his voice heavy with regret. "Your mother used the same route when she tried to escape. Did you really think I wouldn't monitor every passage in the facility?"

Eva's blood went cold. He'd known about her mother's escape route. He'd let Ingrid think she was free, then intercepted her before she could reach the outside world.

"I gave your mother the same choice I'm giving you," Klaus said, his voice carrying through the stone. "Join us willingly, help us create a better world, or be eliminated as a security risk. She chose poorly."

The false wall began to move. Klaus had found the mechanism.

Eva scrambled through the tunnel, her expensive laboratory attire tearing on the rough stone. Behind her, she could hear Klaus's voice becoming more distant but also more threatening.

"Security Team Alpha, seal the forest perimeter. Team Beta, coordinate with local authorities. I want her found within the hour."

The tunnel ended at a concealed grate that opened onto a steep hillside covered in Alpine forest. Eva pushed through the grate and into the night, the December cold hitting her like a physical blow. She was free, but barely.

The monastery blazed with lights behind her, security teams mobilizing with military efficiency. She could hear the vehicles starting, helicopters warming up—the coordinated response of an organization with seemingly unlimited resources.

Eva's phone buzzed with an incoming call. Klaus's name appeared on the screen.

Against her better judgment, she answered.

"Eva." Klaus's voice was calm, almost gentle. "You're making a terrible mistake. You have no idea what you're up against."

"I know exactly what I'm up against," Eva replied, her voice steadier than she felt. "I'm up against my own father, who murdered my mother and is planning to murder four billion people."

"Your mother was weak. She couldn't see past immediate suffering to long-term benefit. I had hoped you would be stronger."

"Stronger?" Eva's voice cracked. "You call genocide strength?"

"I call it necessity." Klaus's tone became harder. "The world is dying, Eva. Overpopulation, genetic degradation, and the collapse of civilized society. The Phoenix Order is offering humanity a chance to survive, to evolve, to become something better than what we are."

"By murdering billions of innocent people?"

"By removing the genetic dead weight that's dragging our species toward extinction." Klaus's voice carried the fervent certainty of a true believer. "The Lethe Virus will eliminate the weak, the inferior, the genetically compromised. What remains will be stronger, purer, more capable of surviving in the world we're creating."

Eva felt sick. This was her father, the man who'd taught her to love science, who'd comforted her nightmares, who'd shown her the beauty of molecular structures and genetic sequences. And he was talking about human beings like they were lab specimens.

"I won't let you do this," she said quietly.

"You can't stop us. The Phoenix Order has resources you can't imagine, connections you can't fathom. We have people in every government, every intelligence agency, every major corporation. You're one woman against a global conspiracy."

"Then I'll find help."

Klaus's laugh was bitter. "From whom? The CIA analyst whose work we've perverted into our operational framework? Dr. Benjamin Carter is either going to join us or be destroyed by his own guilt. Either way, he won't be able to help you."

"We'll see about that."

Eva ended the call and turned off her phone, knowing that keeping it active would allow them to track her location. She was alone in the Alpine forest, pursued by a global conspiracy with unlimited resources, carrying the only evidence that could expose their plan.

But she also carried something else—her mother's hidden research, encoded in a childhood lullaby that held the key to stopping the Lethe Virus. The counter-virus was possible, but Eva needed help to decode it, needed expertise she didn't possess.

She needed Benjamin Carter.

As Eva began her descent through the forest, she could hear the helicopters lifting off behind her, their searchlights sweeping the tree line. The hunt was on, and she was the prey.

But she was also the hunter now. The Phoenix Order had created their perfect weapon, but they'd also created their perfect enemy. Eva Richter, the sheltered scientist who'd been manipulated her entire life, was about to become the most dangerous person in the world.

Because she knew their secret. She had their files. And she was willing to die to stop them.

The Phoenix Order thought they were reshaping humanity's future.

Eva was about to show them that the future could fight back.

The chase through the Alps was just beginning, and she intended to win.

# Chapter 8: The Debate

Ben Carter crouched in the shadows of a Berlin parking garage, watching his apartment building through night-vision binoculars. Four days had passed since Sarah's assassination, four days of careful movement through the city's underground networks, staying one step ahead of the hunters who'd killed his former handler.

The apartment looked normal from the outside—same yellow brick facade, same lighted windows, same elderly neighbor walking her dog. But Ben's trained eye caught the details that marked it as compromised: the white van parked with perfect sight lines to his entrance, the man reading a newspaper at the café who never turned a page, the subtle repositioning of his window blinds that indicated someone had been inside.

His sanctuary was gone.

Ben's phone buzzed with an encrypted message from a number he didn't recognize:

Dr. Carter, my name is Eva Richter. I have information about the Phoenix Order and the people who killed Sarah Chen. We need to meet. Your life depends on it. —E.R.

Ben deleted the message immediately, but his hands were shaking. Eva Richter. The name meant nothing to him, but the timing was suspicious. Four days of silence, then a mysterious woman claiming to have information about Sarah's murder.

Either she was legitimate, or she was bait.

Ben's training screamed at him to run. Disappear into the European underground, activate one of his backup identities, and find a quiet corner of the world where Dr. Benjamin Carter, the mild-mannered professor, could reinvent himself as someone else entirely. It was the smart move, the safe move, the move that would keep him alive.

But his conscience demanded something else entirely.

Sarah had died trying to warn him about something. She'd emerged from retirement, risked her life, faced professional killers—all to deliver information that might save lives. And Ben had let her die while he escaped through a window.

The guilt was eating him alive.

Ben made his decision. He would approach the apartment, assess the situation, and try to recover any intelligence Sarah might have left behind. It was dangerous, possibly suicidal, but it was better than living with the knowledge that her death had been meaningless.

The building's maintenance entrance was hidden in an alley behind the main structure. Ben had chosen this apartment specifically for its multiple access points, its concealed sight lines, and its potential for rapid escape. Now those same features would help him infiltrate his own home.

The lock was sophisticated, but Ben's skills hadn't atrophied during his academic years. Three minutes later, he was inside the building's utility areas, moving through familiar corridors toward his second-floor apartment.

The elevator would be monitored, and the main stairwell would be too exposed. But the building's original architecture included a service stairway that connected to the roof access. Ben climbed carefully, his senses heightened to combat-level awareness.

At the second-floor landing, he found the first sign of trouble. His apartment door was slightly ajar—not enough for a casual observer to notice, but enough to indicate that someone wanted him to know they'd been inside.

Ben drew the Glock from his go-bag, checking the chamber and safety in one smooth motion. His academic persona had been stripped away, replaced by the intelligence operative he'd once been.

The apartment was a disaster zone. Furniture is overturned, books are scattered, and electronics are destroyed. But it wasn't random vandalism—it was a professional search, methodical and thorough. Someone had been looking for something specific.

Ben moved through the rooms carefully, cataloging the damage. His computer was gone, along with his backup drives and encrypted communications equipment. The hidden safe behind his bookshelf had been opened, and its contents had been removed. Even his academic papers had been searched, individual pages scattered across the floor.

But the searchers had missed something. In the kitchen, behind the refrigerator, Ben had hidden a small device that looked like a simple phone charger. It was actually a sophisticated recording system, designed to capture audio and video from anyone who entered the apartment.

Ben retrieved the device and plugged it into his laptop. The footage was revealing:

Three men in expensive suits were moving through his apartment with professional efficiency. They spoke in accented English—Russian, Ben realized, the same language he'd heard from the tactical team that had pursued him through the basement.

The leader was methodical, systematic, and clearly experienced in intelligence gathering. He photographed everything, took samples of DNA evidence, and even collected dust from Ben's bookshelf for analysis.

But it was the conversation that made Ben's blood run cold:

"Carter's psychological profile suggests he'll return here," the leader said. "Emotional attachment to familiar spaces, reluctance to abandon personal possessions. Classic patterns from Damascus-era operatives."

"What about the Richter woman?" asked one of the others. "Von Hess wants her eliminated before she can make contact."

"She's resourceful, but predictable. She'll follow her mother's playbook—try to contact Carter through academic channels, appeal to his sense of justice. We'll intercept her before she can reach him."

Ben's mind raced. Eva Richter was real, and she was in danger. The Phoenix Order was hunting her, using Ben as bait to draw her out.

His phone buzzed again. Another message from Eva:

They found my father's files. 4 billion people will die on Christmas Eve unless we stop them. I know you don't trust me, but I'm the only one who can prove what's coming. Please. —E.R.

Ben stared at the message, his analytical mind processing the implications. Four billion people. Christmas Eve. The scope was almost incomprehensible, but the desperation in Eva's words felt genuine.

His phone buzzed a third time:

I'm outside your building. I know it's compromised. The Phoenix Order killed my mother, just like they killed Sarah Chen. If you're reading this, you have thirty seconds to decide: run and hide, or help me save the world. I'll be in the blue BMW across the street. —E.R.

Ben moved to his window, peering through the blinds. There—a blue BMW sedan, engine running, a woman's silhouette visible in the driver's seat. As he watched, she lifted her phone, obviously texting.

His phone buzzed:

I see the surveillance van. They're closing in on both of us. This is your last chance. —E.R.

Ben's training told him to run. The woman could be Phoenix Order, the car could be a trap, and the entire scenario could be an elaborate deception designed to capture him. But his conscience told him something else.

Sarah had died trying to warn him about a threat he didn't understand. If Eva Richter was legitimate, if she really had information about the Phoenix Order's plans, then running would make him complicit in whatever was coming.

The debate that had been raging in his mind for four days crystallized into a single moment of choice. He could disappear, save himself, and live with the knowledge that he'd abandoned his responsibilities again. Or he could take the risk, trust a stranger, and possibly prevent another catastrophe.

Ben thought about his students, their earnest faces, and the questions they had about moral courage. How could he teach them about the importance of taking a stand while hiding from his own responsibilities?

He made his decision.

Ben grabbed his go-bag and headed for the window. The fire escape was exposed, but the BMW was positioned to provide cover from the surveillance van. If Eva were legitimate, she'd chosen her position well.

If she were Phoenix Order, he was walking into a trap.

As he climbed down the fire escape, Ben's phone exploded with activity. Multiple messages, all from Eva:

Hurry. They're moving.

Van just called for backup.

Two more cars are approaching from the south.

Ben, if you're coming, come now.

Ben reached the street level and sprinted toward the BMW. Through the windshield, he could see Eva Richter clearly for the first time—young, blonde, beautiful, and absolutely terrified. Her hands were shaking as she gripped the steering wheel, her eyes darting between Ben and the surveillance van.

He was ten feet from the car when the van's doors opened and tactical teams poured out. Professional killers, armed and coordinated, moving with lethal precision.

Ben dove into the BMW just as the first shots shattered the rear window. Eva floored the accelerator, tires screaming against asphalt as they careened through Berlin's narrow streets.

"Drive!" Ben shouted, unnecessary advice since Eva was already pushing the BMW to its limits.

"I'm Dr. Eva Richter," she said, her voice steady despite the chaos. "My father is Klaus Richter, and he's planning to kill four billion people with a bioweapon called the Lethe Virus."

Ben checked the side mirror. Three vehicles were pursuing them, their coordination suggesting professional training and unlimited resources.

"Your father is in the Phoenix Order?"

"He's one of their leaders. And they're using your analytical methods to plan the attack." Eva took a sharp turn, the BMW's tires protesting but holding. "Ben, I need your help. I have the evidence to stop them, but I can't do it alone."

The debate was over. Ben had chosen action over safety, engagement over exile. The Phoenix Order had forced him to pick a side, and he'd finally made his choice.

"Where are we going?" he asked.

"Safe house. I'll explain everything, but we need to get out of Berlin. They have resources we can't imagine."

As they drove through the night, pursued by professional killers and carrying the weight of an impossible story, Ben realized that his academic life was truly over. The Phoenix Order had awakened something they probably wished they'd left sleeping.

Dr. Benjamin Carter, the mild-mannered professor, was gone.

The intelligence operative was back.

And he was about to remind the Phoenix Order why he'd once been considered one of the CIA's most dangerous assets.

The hunt was on, and Ben Carter was no longer just the prey.

He was ready to become the hunter.

But first, he needed to understand exactly what they were hunting.

And the terrified woman beside him, driving through Berlin's streets like her life depended on it, might be the only person who could provide those answers.

The Phoenix Order had made its first mistake.

They'd given Ben Carter a reason to fight back.

Now he just had to make sure they both survived long enough to stop whatever was coming on Christmas Eve.

The debate was over.

The war was just beginning.

# Chapter 9: First Contact

The Hotel Adlon's bar was a temple to old-world elegance, its marble columns and crystal chandeliers evoking the grandeur of imperial Berlin. Ben Carter sat at the far end of the mahogany bar, nursing a glass of whiskey and trying to blend into the crowd of well-dressed tourists and business travelers. Four days had passed since Sarah's assassination, four days of careful movement through Berlin's shadows, staying one step ahead of the hunters who'd killed his former handler.

The encrypted message from Eva Richter had arrived that morning, redirecting their planned meeting from the original location to this more public venue. Ben had almost ignored it—after Sarah's death, he trusted no one. But something about the message's tone, its desperate urgency, had convinced him to take the risk.

They know about the original meeting point. Hotel Adlon bar, 9 PM. I'll be wearing a blue dress, carrying a red purse. I have the information that led to your handler's death. If you don't come, millions will die. Please. I'm running out of time. —E.R.

Ben checked his watch: 9:17 PM. Either Eva Richter was very late, very careful, or very dead. He'd positioned himself with clear sight lines to all entrances, his back to the wall, ready to move at the first sign of trouble. The bar's ambient noise—conversations in multiple languages, clinking glasses, soft piano music—would mask any urgent movement.

He was scanning the room for the third time when he saw her.

She entered through the main lobby, and Ben's breath caught. Eva Richter was stunning in a way that made his analytical mind immediately suspicious. Mid-twenties, blonde hair pulled back in an elegant chignon, wearing a blue dress that probably cost more than Ben's monthly salary. She carried herself with the confidence of someone accustomed to male attention. Still, there was something else—a tension in her shoulders, a careful way of surveying the room that suggested she was as paranoid as he was.

Their eyes met across the crowded bar, and Ben saw something that made his chest tighten. Terror. Barely controlled, carefully hidden, but unmistakable to someone who'd spent years reading faces and body language.

Eva made her way to the bar, ordering a martini with the kind of nervous precision that suggested she rarely drank. When she sat down two stools away from Ben, he could see her hands shaking slightly.

"Dr. Carter?" Her voice was soft, accented, cultured. German, but with the precise diction of someone who'd learned English in expensive schools.

"You're Eva Richter."

"Yes." She took a large sip of her martini, wincing slightly at the alcohol. "Thank you for coming. I wasn't sure you would."

Ben studied her face, looking for deception, for the tells that would indicate she was working for the Phoenix Order. But all he saw was fear and desperation. "Your message mentioned information about my handler's death."

"Sarah Chen. She was investigating the Phoenix Order, trying to understand their connection to you. They killed her to prevent her from warning you about what's coming."

"What's coming?"

Eva's laugh was bitter, almost hysterical. "The end of the world. Disguised as a pandemic." She finished her martini in one gulp, then signaled for another. "Dr. Carter, how familiar are you with genetic engineering?"

"Academic level only. I'm a historian, not a scientist."

"Then let me give you a crash course in genocide." Eva's second martini arrived, and she immediately took a large sip. "Three years ago, my father recruited me to work on what I thought was revolutionary medical research. Genetic therapies are targeted treatments for inherited diseases. Noble work, life-saving work."

Ben watched her carefully. The alcohol was loosening her tongue, but her story had the ring of truth. "What were you actually working on?"

"A bioweapon. The most sophisticated viral pathogen ever created." Eva's voice dropped to barely above a whisper. "It's called the Lethe Virus. It targets specific genetic markers and allows for population-selective elimination. Kill the undesirable, spare the chosen. The perfect tool for racial purification."

Ben's blood went cold. "You're talking about targeted genocide."

"I'm talking about the systematic murder of four billion people." Eva's eyes were bright with unshed tears. "They're planning to release it simultaneously at twelve major airports on Christmas Eve. Global distribution, maximum impact, perfect timing during the holiday travel season."

The implications hit Ben like a physical blow. Christmas Eve was eight days away. "Who's 'they'?"

"The Phoenix Order. A neo-Nazi organization that's been planning this for decades. They've infiltrated governments, scientific institutions, and pharmaceutical companies. They have unlimited resources and government protection."

Ben's analytical mind began processing the information, looking for inconsistencies and logical flaws that would indicate fabrication. But Eva's story was horrifyingly plausible. "Where do I fit in?"

"You're the patsy." Eva's voice was steady now, the alcohol giving her courage. "They're going to frame you as the mastermind. Your expertise in population analysis, your knowledge of target identification, your history with the Syrian operation—it all makes you the perfect fall guy."

"How do you know about Syria?"

"Because they have your classified files. Complete dossiers on your intelligence work, your psychological profile, and your personal weaknesses. They know exactly how to manipulate you." Eva leaned closer, her voice urgent. "Dr. Carter, they're using your analytical methods to plan the attack. Your protocols for target selection, your frameworks for acceptable casualty rates. They're turning your own work against you."

Ben felt the familiar weight of guilt settling on his shoulders. Damascus all over again, but on a global scale. "Why should I believe you?"

"Because," Eva said, reaching into her purse, "I have proof."

She placed a small data drive on the bar between them. "Everything's on there. The virus specifications, the target list, and the release schedule. The Phoenix Order's complete operational plan."

Ben stared at the drive, his mind racing. "If you have this information, why not take it to the authorities?"

"What authorities?" Eva's laugh was bitter. "The Phoenix Order has people everywhere. Government officials, intelligence agencies, and military commanders. Anyone I trust could be working for them."

"Then why trust me?"

"Because you're the only one who can stop them." Eva's eyes met his, and Ben saw something that made his chest tighten. Hope. Desperate, fragile, but genuine. "My mother was working on a counter-virus before they killed her. She hid the formula, but I need help to decode it. I need your analytical skills to understand how they're planning to deploy the weapon."

Ben's whiskey sat untouched, forgotten. "Your mother was murdered?"

"By my father. Klaus Richter, the brilliant scientist who raised me, and taught me to love learning and discovery. He killed her because she threatened to expose their plan." Eva's voice broke slightly. "I found her research notes. She was trying to contact intelligence analysts, people who could help her understand the Phoenix Order's operational methods."

"She was trying to contact me?"

"Your name was in her files. She believed your expertise in population targeting could help develop countermeasures." Eva's hand moved across the bar, almost touching Ben's. "Dr. Carter, I know this is impossible to believe. I know it sounds like the ravings of a paranoid conspiracy theorist. But I'm begging you to look at the evidence."

Ben studied her face, looking for the tells that would indicate deception. But all he saw was a brilliant woman destroyed by the discovery of her father's evil, reaching out to a stranger because she had no one else left to trust.

"The data drive," he said quietly. "What's on it?"

"Everything. Virus specifications, target locations, personnel files, and financial records. Enough evidence to expose the entire conspiracy."

Ben's hand moved toward the drive, then stopped. "If I take this, I'm committing myself to a course of action that could get us both killed."

"If you don't take it, billions of people will die."

Before Ben could respond, Eva's eyes widened in terror. She was looking over his shoulder, toward the bar's entrance.

"They found us," she whispered.

Ben turned to see three men in expensive suits moving through the crowd with practiced precision. They weren't obviously armed, but their movement patterns, their coordination, their systematic scan of the bar—everything screamed professional operatives.

"How?" Ben asked.

"I don't know. I was so careful, I checked for surveillance, I—" Eva's voice cracked. "Oh God, they're going to kill us both."

The lead operative had spotted them. He spoke into a concealed microphone, and Ben saw two more men entering through the hotel's side entrance. They were being systematically surrounded.

"Eva," Ben said quietly, "do you trust me?"

"I don't have a choice."

"When I move, you follow. Stay close, do exactly what I say, and we might survive the next five minutes."

The lead operative was twenty feet away, his hand moving inside his jacket. Ben had seconds to act.

He grabbed the data drive, pocketing it in one smooth motion. Then he took Eva's hand, pulling her toward the bar's service entrance.

"Move. Now."

The first shot shattered the mirror behind the bar, sending crystal fragments cascading onto the polished wood. The bar erupted in chaos—screaming patrons, overturned tables, the acrid smell of gunpowder mixing with spilled alcohol.

Ben pulled Eva behind the bar, using the solid mahogany as cover. His mind was racing, calculating angles, exit routes, tactical options. The Phoenix Order operatives were professionals, but they were also constrained by the public setting.

"Can you run?" he asked Eva.

"Yes."

"Then run."

They moved through the bar's service area, Ben's dormant training awakening with each step. Behind them, he could hear the operatives coordinating their pursuit, their professional calm more terrifying than any shouted threats.

The hotel's kitchen was a maze of stainless steel and steam, busy with the dinner service. Ben and Eva moved through it like ghosts, past confused chefs, and startled waiters.

"The service elevator," Eva pointed to a small lift at the far end of the kitchen. "It goes to the parking garage."

As they ran, Ben's mind processed what had just happened. Eva's story was either genuine or the most sophisticated deception he'd ever encountered. But the Phoenix Order's response—immediate, violent, professional—suggested she was telling the truth.

In the elevator, as they descended toward the garage, Ben looked at Eva's face. She was terrified but determined, her elegant facade stripped away to reveal someone who'd lost everything and was fighting for humanity's survival.

"Eva," he said quietly, "if we survive this, I'm going to need you to tell me everything. Every detail, every piece of evidence, every connection you can remember."

"Does that mean you believe me?"

Ben thought about the operatives, their professional precision, their willingness to kill in a public place. He thought about Sarah's murder, the Syrian operation, the pattern of manipulation and violence that seemed to follow him wherever he went.

"It means I'm tired of running," he said. "If the Phoenix Order wants a war, they're going to get one."

The elevator opened onto the parking garage, and Ben pulled Eva into the shadows between the cars. They had minutes, maybe less, before the operatives reached them.

"Your car?" he asked.

"BMW, blue, section C."

As they moved through the garage, Ben realized that his academic life was truly over. The Phoenix Order had forced him to choose between hiding and fighting, between safety and responsibility.

For the first time in three years, he was ready to choose action over safety.

The Phoenix Order had awakened something they probably wished they'd left sleeping.

And Ben Carter, former CIA analyst, was about to remind them why he'd once been considered one of the agency's most dangerous assets.

# Chapter 10: Into the Fire

The Hotel Adlon's bar had been a sanctuary of old-world elegance just moments before—crystal chandeliers casting warm light over marble columns, the soft murmur of conversation mixing with piano music. Now it was a battlefield.

Ben Carter moved with lethal precision through the chaos, his academic facade stripped away to reveal something far more dangerous. The mild-mannered professor was gone, replaced by a CIA operative whose dormant skills were awakening with each heartbeat.

"Stay behind me," Ben commanded, his voice carrying an authority Eva had never heard before. The transformation was startling—gone was the hesitant historian, replaced by a predator who moved through violence like it was his natural element.

Eva pressed close to Ben's back, her heart hammering as she watched him navigate the bar's overturned tables and scattered glass. She could feel the heat radiating from his body, smell his cologne mixed with adrenaline and something else—something primal and masculine that made her breath catch despite the mortal danger surrounding them.

The first Phoenix Order operative appeared through the smoke, his weapon drawn, eyes scanning for targets. Ben moved faster than Eva thought possible, his hand shooting out to grab a crystal decanter from the bar. The heavy glass caught the operative in the temple, dropping him instantly.

"Jesus," Eva breathed, staring at Ben's fluid, economical violence. "You're not just a professor."

"No," Ben replied, appropriating the fallen operative's weapon with practiced ease. "I'm not."

He checked the chamber, engaged the safety, and tucked the gun into his waistband—movements so smooth they seemed choreographed. Eva found herself transfixed by the transformation, by the deadly competence he'd been hiding beneath his academic exterior.

"We need to move," Ben said, his eyes scanning the bar for threats. "They'll have the exits covered, backup teams in position. This is a coordinated extraction."

"Extraction?" Eva's voice was barely a whisper.

"They want us alive. For now." Ben's hand found hers, and Eva felt an electric shock at the contact. His fingers were strong, calloused, nothing like the soft hands she'd expected from a university professor. "Can you run?"

Eva nodded, not trusting her voice. The warmth of his touch was spreading up her arm, creating a sensation that had nothing to do with fear and everything to do with the way Ben's protective instincts had awakened something deep inside her.

They moved through the bar's chaos, Ben's body shielding Eva from the worst of the violence. She could feel the controlled power in his movements, the way he positioned himself between her and danger without conscious thought. It was protective, primal, and utterly intoxicating.

The hotel's lobby was a maze of panicked guests and scattered luggage. Ben navigated it like a predator, his trained eye cataloging threats and opportunities. Eva stayed close, her hands occasionally brushing against his back, feeling the tense muscles beneath his jacket.

"There," Ben pointed to a service corridor behind the concierge desk. "Staff areas will have multiple exits."

As they moved toward the corridor, Eva caught sight of herself in a mirror—disheveled, wild-eyed, her elegant blue dress torn and stained. She looked like a woman who'd been through hell.

But when she looked at Ben, she saw something else entirely. Purpose. Deadly competence. The kind of controlled violence that should have terrified her, but instead sent heat pooling low in her belly.

"Ben," she said, her voice breathy. "How did you—"

"Later," he interrupted, his hand finding the small of her back to guide her forward. The contact sent shivers down Eva's entire body. "Right now, we survive."

The service corridor was narrow, forcing them to move in single file. Eva was acutely aware of Ben behind her, his presence solid and protective. When they reached a junction, he moved past her to check the route ahead, his body brushing against hers in the confined space.

The contact was electric. Eva's breath caught as Ben's chest pressed against her back, his arms reaching around her to test the door handle. She could feel his heartbeat, steady and strong, and smell the masculine scent of his cologne, mixed with something darker and more dangerous.

"Clear," Ben whispered, his breath warm against her ear.

Eva turned in the narrow space, finding herself face-to-face with Ben, their bodies inches apart. His eyes were intense, predatory, but she could see something else there—concern for her safety, protectiveness that went beyond mere partnership.

"You're bleeding," she said, noticing a cut on his cheek from the bar fight.

Without thinking, Eva reached up to touch the wound. Ben's eyes darkened as her fingers made contact with his skin, and Eva felt the world narrow to just the two of them, the danger around them fading into background noise.

"Eva," Ben's voice was rough, strained.

"I know," she whispered, her thumb tracing the edge of the cut. "I know this is crazy, but—"

The sound of approaching footsteps shattered the moment. Ben's training kicked in immediately, his body tensing as he positioned himself between Eva and the threat.

"Move," he commanded, his voice once again carrying that deadly authority.

They emerged from the service corridor into the hotel's kitchen, a maze of stainless steel and steam. The dinner service had been abandoned, leaving the space eerily quiet except for the hiss of gas burners and the distant sound of chaos from the lobby.

Ben moved through the kitchen like a ghost, his military training evident in every step. Eva followed, mesmerized by the way he seemed to anticipate every angle, every potential threat.

"There," Ben pointed to a loading dock at the far end of the kitchen. "Service entrance. If we can reach it—"

The first shot shattered a hanging pot rack, causing cookware to cascade to the floor. Ben dove, tackling Eva behind the prep counter as more shots rang out.

The impact drove them both to the floor, Ben's body covering hers protectively. Eva found herself trapped beneath him, acutely aware of his weight, his strength, the way his muscles tensed as he prepared to move.

"Are you hurt?" Ben's voice was urgent, his hands running over her body to check for wounds.

The contact was intimate, necessary, but it sent fire through Eva's veins. His hands were skilled, gentle despite the violence surrounding them, and she found herself arching slightly into his touch.

"I'm fine," she managed, though her voice was breathy with something that had nothing to do with fear.

Ben's eyes met hers, and for a moment, the world stopped. They were pressed together on the kitchen floor, danger surrounding them, but all Eva could think about was the way Ben's protective instincts had awakened something primal in her response.

"We need to move," Ben said, but his voice lacked conviction. His eyes were fixed on her lips, and Eva felt her breath catch.

"Ben," she whispered, her hand finding his chest.

The moment stretched between them, electric and dangerous. Eva could feel Ben's heartbeat beneath her palm, could see the war between duty and desire playing out in his eyes.

Another shot forced them back to reality. Ben rolled off her, his movements fluid and professional, but Eva caught the slight tremor in his hands as he checked his weapon.

"The loading dock," Ben said, his voice steady despite the tension crackling between them. "We go fast, we go quiet."

They moved through the kitchen, using the prep stations as cover. Ben's tactical awareness was extraordinary, but Eva noticed that he never let her get more than an arm's reach away; his protective instincts overrode his training.

At the loading dock, Ben checked the door's security system with practiced efficiency. "It's locked, but I can—"

"No need," Eva interrupted, pulling a keycard from her jacket. "Hotel security card. I lifted it from the concierge desk."

Ben stared at her, something like admiration flickering in his eyes. "You're full of surprises."

"So are you, Professor Carter." Eva's voice carried a hint of challenge, of invitation.

The loading dock opened onto a narrow alley behind the hotel. Ben checked both directions, his movements economical and professional. But when he reached for Eva's hand to guide her forward, the contact sent sparks through both of them.

"There," Ben pointed to a BMW sedan parked at the alley's mouth. "If we can reach it—"

"That's my car," Eva said, surprising him again.

"You came prepared."

"I came desperate." Eva's honesty was raw, vulnerable. "Ben, I need you to know—if we don't survive this, I'm glad I found you. I'm glad I'm not facing this alone."

Ben's expression softened, the deadly operative momentarily replaced by the man beneath. "Eva, you're not alone. Whatever happens, we're in this together."

The promise hung between them as they moved toward the car. Eva could feel the weight of it, the implication that went beyond mere survival. Ben had chosen her, chosen to fight alongside her, chosen to risk everything for a woman he'd known for less than an hour.

But as they reached the BMW, Eva realized that she'd made the same choice. She was trusting her life, her mission, her heart to a man who'd been a stranger thirty minutes ago.

The car's engine roared to life, and Ben took the wheel with the same deadly competence he'd shown in the bar. But as they pulled away from the hotel, Eva noticed that his free hand found hers, their fingers intertwining with an intimacy that spoke of more than tactical partnership.

"Where?" Ben asked, his voice rough with adrenaline and something else.

"Safe house," Eva replied, her thumb tracing circles on his palm. "I know a place."

As they drove through Berlin's darkened streets, Eva was acutely aware of Ben beside her—his controlled strength, his protective instincts, the way he'd risked everything to save her. The elegant professor was gone, replaced by a warrior who'd chosen to fight for her.

The hotel attack had changed everything. They were no longer strangers brought together by circumstance. They were partners, allies, two people who'd seen each other's true nature under fire and chosen to stand together.

But Eva could feel something else building between them, something that went beyond partnership or shared danger. The way Ben's eyes had darkened when she'd touched his face, the way her body had responded to his protective embrace, the electric current that passed between them every time they touched.

They were fugitives now, hunted by a global conspiracy with unlimited resources. But they were also something else—two people who'd found each other in the darkness, who'd chosen to fight together against impossible odds.

The Phoenix Order had made a critical mistake. They'd assumed that forcing Ben and Eva together would make them easier to capture.

Instead, they'd created something far more dangerous than either individual could have been alone.

They'd created a partnership born in violence, forged in trust, and tempered by the kind of attraction that could move mountains.

The war was just beginning, but Ben and Eva were no longer fighting it alone.

They were fighting it together.

And that made all the difference in the world.

The Phoenix Order was about to learn that some bonds couldn't be broken by bullets or fear.

Some bonds were stronger than conspiracy, more dangerous than any weapon.

Some bonds were forged in fire and baptized in blood.

And those bonds would be their undoing.

# Chapter 11: The Christmas Market Chase

The BMW's engine screamed as Eva floored the accelerator, tires smoking against the wet Berlin asphalt. In the rearview mirror, Ben could see three black SUVs maintaining pursuit, their headlights cutting through the December night like predatory eyes.

"They're coordinating," Ben shouted over the engine noise, his trained eye cataloging the pursuit vehicles' movements. "Professional formation, radio communication. This isn't random—they're herding us."

Eva's knuckles were white on the steering wheel as she took a sharp right onto Unter den Linden. "Herding us where?"

"Somewhere with fewer witnesses and more firepower." Ben checked his side mirror as one of the SUVs tried to ram them from behind. "They want us alive, or they would have taken us out already."

The Christmas market at Gendarmenmarkt appeared ahead of them, its warm lights and festive crowds creating an island of normalcy in their nightmare. Hundreds of people strolled between wooden stalls, drinking mulled wine and shopping for presents, utterly unaware that death was racing toward them at sixty miles per hour.

"Eva, you need to slow down. We can't—"

"I know." Her voice was tight with concentration. "But if we stop, they'll kill us both."

Ben's mind raced through tactical options. The market was crowded, chaotic, perfect for losing pursuers but also dangerous for civilians. The Phoenix Order had already shown they were willing to kill innocents—the Hotel Adlon attack had proven that. However, running through a Christmas market at high speed would likely result in casualties.

"There," Eva pointed to a narrow street that ran parallel to the market. "We can use the crowd as cover, but we need to abandon the car."

Ben nodded, understanding her logic. The BMW was fast but conspicuous. In the market's chaos, they could disappear among the shoppers—if they could get there alive.

Eva slammed on the brakes, the BMW skidding to a stop at the market's edge. The pursuing SUVs were seconds behind, their drivers having no choice but to follow suit or risk a massacre.

"Run," Ben said, grabbing Eva's hand as they bolted from the car.

The Christmas market exploded into chaos as they sprinted through the crowd. Behind them, Ben could hear car doors slamming, the sound of boots on cobblestones, and shouted commands in multiple languages. The Phoenix Order was deploying on foot, their pursuit becoming a manhunt.

Eva pulled Ben toward a cluster of wooden stalls selling handmade crafts. "This way. I know Berlin's underground."

They dove between the stalls, using the maze of wooden structures as cover. The scent of cinnamon and roasted almonds filled the air, a grotesque normality against the backdrop of mortal danger.

"There!" A voice shouted behind them. "Northwest corner, moving toward the cathedral!"

Ben risked a glance back and saw something that made his blood freeze. The pursuing operatives weren't just following them—they were coordinating through sophisticated communication devices, their movements too precise for simple radio chatter.

"Eva, they're using real-time tracking. Facial recognition, probably AI-assisted."

"What?" Eva's eyes widened as she processed the implications. "That's impossible. The processing power required—"

"The Phoenix Order has government backing. Military-grade surveillance." Ben pulled her behind a larger stall selling Christmas ornaments. "We need to change our appearance, now."

Eva's scientific mind kicked into overdrive. "The facial recognition algorithms focus on bone structure, eye spacing, and distinctive features. If we can alter our silhouettes..."

She grabbed a wool hat from a nearby stall, pulling it low over Ben's forehead. He did the same for her, adding a thick scarf that obscured the lower half of her face.

"Better, but not enough," Ben muttered. "They'll adapt."

As if to confirm his words, Ben heard one of the operatives speaking into his device: "Targets have modified appearance. Adjusting parameters for thermal imaging and movement patterns."

"Thermal imaging?" Eva's voice was barely a whisper.

"They're not just tracking our faces. They're tracking our heat signatures, our movement patterns, probably our heart rates." Ben's mind raced through countermeasures. "We need to get underground, somewhere with multiple heat sources and escape routes."

"The U-Bahn station," Eva said. "Gendarmenmarkt has direct access to the underground system."

They moved through the market, using the crowd as cover while staying alert for the operatives hunting them. Ben's training allowed him to spot the hunters—their too-calm movement through the chaos, their coordinated positioning, their subtle hand signals.

There were at least eight of them, probably more. All professionals, all armed, all focused on capturing or killing two people in a crowd of hundreds.

"Movement, southeast quadrant," crackled a voice from nearby. "Targets attempting to reach subway access."

Ben grabbed Eva's arm. "They're anticipating our route. We need to—"

The shot came from behind them, a suppressed rifle that sent a bullet whistling past Ben's ear. The Christmas ornament stall exploded in a shower of glass and tinsel.

Screaming erupted from the crowd. Shoppers scattered in all directions, their festive evening transformed into a nightmare of running figures and broken glass.

"Down!" Ben tackled Eva behind a large wooden stall, his body shielding her from the chaos. Around them, the market devolved into panic as more shots rang out.

"They're not trying to avoid civilian casualties," Eva said, her voice tight with horror.

"No, they're using the panic to flush us out." Ben peered around the edge of the stall, cataloging the positions of the operatives. "They want us to run toward the subway, where they can control the environment."

Another shot shattered the stall's wooden frame, sending splinters flying. Ben could hear the operatives moving, their boots heavy on the cobblestones as they closed in.

"There," Eva pointed to a drainage grate twenty feet away. "Berlin's storm drain system connects to the subway tunnels. If we can reach it..."

"Too exposed." Ben calculated angles and distances. "We need a distraction."

Eva's scientific mind provided the solution. "The mulled wine stand. The propane tanks."

Ben followed her gaze and understood immediately. The traditional German Feuerzangenbowle stand used open flames to heat the wine, the propane tanks clearly visible beneath the wooden counter.

"Can you hit it from here?" Eva asked.

Ben smiled grimly. "I'm a historian, not a marksman. But I spent three years in tactical training."

He aimed carefully, his academic persona stripped away completely. The shot was perfect—the propane tank exploded in a ball of flame that sent burning wine cascading across the cobblestones.

The explosion created perfect chaos. Screaming shoppers, emergency sirens, smoke, and flames that would confuse thermal imaging. In the confusion, Ben and Eva sprinted toward the drainage grate.

Ben's fingers found the heavy iron cover, muscles straining as he lifted it. "Go!"

Eva dropped into the storm drain, her elegant dress immediately soaked by the frigid water below. Ben followed, pulling the grate closed behind them just as the first operative reached their position.

The storm drain was a concrete tunnel barely wide enough for two people, filled with rushing water that came up to their knees. The sound of their movement echoed off the walls, but it was better than being tracked by thermal imaging.

"This way," Eva said, her voice echoing in the confined space. "The tunnel connects to the subway system about two hundred meters east."

They moved through the darkness, guided by Eva's knowledge of Berlin's underground infrastructure. Behind them, Ben could hear the operatives attempting to follow, their progress slowed by the tunnel's narrow confines.

"They're still coming," Ben said, noting the flashlight beams reflecting off the water behind them.

"I know." Eva's voice was steady despite their desperate situation. "But I have an idea."

She led him to a junction where three tunnels met, the water level deeper and the current stronger. "The central tunnel leads to the Spree River. If we can reach it, we can use the current to escape."

Ben nodded, understanding her logic. The river would carry them away from their pursuers, providing multiple escape routes throughout the city.

But as they moved toward the central tunnel, Eva suddenly came to a halt.

"Ben, look."

Ahead of them, red laser dots danced across the tunnel walls. The operatives had somehow gotten ahead of them, anticipating their route and positioning themselves at the junction.

They were trapped.

"Dr. Carter, Dr. Richter," a voice called out, echoing through the tunnels. "There's no need for this to continue. Surrender now, and we can discuss terms."

Ben recognized the voice—the same operative who'd coordinated the Hotel Adlon attack. Professional, calm, utterly without mercy.

"What terms?" Ben called back, buying time while his mind raced through options.

"You join the Phoenix Order willingly and contribute your skills to the cause. Dr. Richter returns to her research, completes her work on the viral delivery system. Everyone benefits."

"And if we refuse?"

"Then you die here, in the tunnels, like the rats you've become."

Ben felt Eva's hand grip his arm. When he looked at her, he saw not fear but determination.

"Ben," she whispered, "I saved some of the thermal charges from the laboratory. If I can create a localized explosion..."

"Too dangerous. You could be killed."

"Better than letting them complete their genocide." Eva's eyes met his, and Ben saw the same moral courage he'd been teaching his students about. "I won't let my father's evil succeed."

Before Ben could stop her, Eva was moving. She pulled something from her jacket—a small device that looked like a modified laboratory tool. Her fingers worked quickly, setting what appeared to be a timer.

"Eva, no—"

The explosion shook the tunnel system, causing water to cascade from the ceiling. The operatives' shouts turned to screams as sections of the tunnel collapsed, blocking their advance.

But Eva was caught in the blast. Ben saw her thrown against the tunnel wall, her body crumpling as debris rained down around her.

"Eva!" Ben fought through the churning water, reaching her just as she began to sink.

She was unconscious, blood streaming from a head wound. Still, she was alive. Ben lifted her in his arms, feeling the weight of her sacrifice, the courage she'd shown in risking everything to save him.

The tunnel behind them was blocked, but the route to the river was clear. Ben carried Eva through the rushing water, her scientific brilliance and moral courage having provided their escape.

As they reached the Spree River and the promise of freedom, Ben realized that Eva Richter had just proven herself to be more than a partner.

She'd proven herself as a hero.

The Phoenix Order had made a crucial mistake. They'd assumed that two academics would be easy prey, that scholars would fold under pressure.

Instead, they'd awakened something far more dangerous than they'd imagined.

Two people who had nothing left to lose, everything to fight for, and the skills to make the Phoenix Order pay for their arrogance.

The chase was over.

The war was just beginning.

And Ben Carter was no longer debating whether to fight.

He was ready to win.

# Chapter 12: Safe Houses and Secrets

The safe house was hidden in plain sight—a modest apartment in Kreuzberg that looked like any other residence in the working-class neighborhood. But as Ben guided Eva through the reinforced door, she realized that nothing about this place was ordinary.

The windows were bulletproof, disguised to look like standard glass. Motion sensors covered every approach angle. The walls were lined with sound-dampening material hidden behind conventional drywall. And in the corner, barely visible unless you knew what to look for, was a sophisticated surveillance system that monitored a six-block radius.

"Jesus," Eva breathed, taking in the paranoid perfection of Ben's preparations. "How long have you been planning for this?"

"Three years," Ben replied, engaging multiple locks on the door. "Since the day I walked away from the agency. I told myself I was being paranoid, but..."

"But paranoia keeps you alive." Eva's scientific mind appreciated the thoroughness of his preparations. "This is impressive, Ben. And terrifying."

Ben moved through the apartment with practiced efficiency, checking the systems and confirming their security perimeter. But Eva noticed the way his shoulders remained tense, the way his eyes never quite stopped scanning for threats.

"We're safe here," he said, finally turning to face her. "For now."

Eva nodded, but she could see the cost of constant vigilance written in the lines around Ben's eyes. Three years of living like this, three years of expecting death to come through the door at any moment.

"Ben," she said softly, "you're bleeding."

He looked down at his hands, seeming surprised by the cuts and scratches from their escape. "It's nothing."

"It's not nothing." Eva moved closer, her voice carrying the authority of someone accustomed to being obeyed in medical situations. "Sit down. Let me look at those wounds."

Ben hesitated, and Eva realized that allowing someone to tend to his injuries required a level of trust he hadn't extended to anyone in years.

"Please," she said, her voice gentler. "I know what I'm doing."

Ben sat on the apartment's simple couch, his body rigid with tension. Eva fetched the first aid kit from the bathroom—military-grade supplies, she noted, not the basic bandages most people kept at home.

"This might sting," Eva warned, settling beside him with antiseptic and gauze.

She began cleaning the cuts on his hands, her touch gentle but efficient. Ben's breathing changed as she worked, becoming deeper, more controlled. Eva could feel the heat radiating from his body, could sense the way he was fighting to remain still under her ministrations.

"You're very good at this," Ben said, his voice rough.

"I've had practice." Eva's fingers traced the edge of an intense cut on his knuckle. "My mother used to patch me up after I'd fallen off my bike or crashed through a laboratory experiment. She always said that healing was just another form of chemistry."

Ben's eyes met hers, and Eva saw something vulnerable there, something that had nothing to do with physical pain.

"Tell me about her," he said quietly.

Eva's hands stilled on his. "She was brilliant. Funny. She saw the best in everyone, even when they didn't deserve it." Her voice caught slightly. "She died trying to save the world from my father's madness. And I was too blind to see what was happening."

"That's not your fault."

"Isn't it?" Eva's eyes filled with tears. "I helped him create the delivery system for the virus. My research, my work—it's all being used to kill billions of people. How do I live with that?"

Ben's free hand came up to cup her face, his thumb brushing away a tear. "By stopping him. By using your knowledge to save the people he wants to kill."

The touch was electric, sending shivers through Eva's entire body. She leaned into his palm, her eyes closing as she savored the warmth of human contact she'd been denied for so long.

"Ben," she whispered, "I'm scared."

"So am I," he admitted, his honesty raw and vulnerable. "I haven't been truly close to anyone since Damascus. I've been hiding, running, pretending that isolation was safety."

Eva opened her eyes, studying his face. "What happened in Damascus?"

Ben's expression darkened, and for a moment, Eva thought he wouldn't answer. Then, slowly, he began to speak.

"I was an intelligence analyst. My job was to identify targets, assess threats, and recommend action. I was good at it—too good." His voice carried the weight of three years of guilt. "There was a terrorist cell planning an attack on a refugee camp. Children, families, and innocent people. I had forty-eight hours to provide actionable intelligence."

Eva continued cleaning his wounds as he spoke, her touch gentle and encouraging.

"I found them," Ben continued. "GPS coordinates, behavioral patterns, communication intercepts. Everything pointed to a compound outside Damascus. I recommended immediate action."

"And?"

"The drone strike was perfect. Eliminated the entire terrorist cell." Ben's voice broke slightly. "It also killed thirty-seven civilians. A wedding party that intelligence had missed. Thirty-seven people who died because I was thorough, precise, and completely wrong about what was happening on the ground."

Eva's heart clenched at the pain in his voice. "Ben, that wasn't your fault. You were trying to save lives."

"I was trying to play God," he replied bitterly. "Making decisions about who lived and who died based on data points and probability matrices. I convinced myself that casualties were acceptable if they prevented greater loss of life."

Eva set down the antiseptic and moved closer, her hands framing his face. "Look at me," she said firmly. "You made a decision based on the information you had. You were trying to save innocent people."

"I was trying to ease my conscience by quantifying human suffering." Ben's eyes met hers, and Eva saw the self-loathing that had been eating him alive for three years. "The Phoenix Order is using my methodology, my frameworks for acceptable casualties. They're turning my work into justification for genocide."

"Then we stop them," Eva said with fierce determination. "We use your knowledge, my research, everything we've learned to make sure they never succeed."

Ben's hands covered hers, and Eva felt the calluses on his palms, the strength in his fingers. "Eva, you don't understand. I'm not a hero. I'm a man who's been running from his mistakes for three years."

"Maybe," Eva said, her voice soft but steady. "But you're also a man who risked everything to save me tonight. A man who chose to fight when he could have run."

She leaned closer, her forehead touching his. "Ben, we're both carrying guilt we don't deserve. But we're also the only ones who can stop this."

The space between them seemed to shrink, charged with electricity that had nothing to do with the danger they'd escaped. Eva could feel Ben's breath on her lips, could see the war between desire and duty playing out in his eyes.

"Eva," he said, his voice rough with want. "This is crazy. We barely know each other."

"I know enough," she replied, her hands sliding down to rest on his chest. "I know you're brave and kind, and that you've been punishing yourself for something that wasn't your fault. I know that when those men were shooting at us, you didn't hesitate to put yourself between me and danger."

Ben's hands moved to her waist, pulling her closer. "I know you're brilliant, and courageous, and that you've sacrificed everything to stop your father's madness. I know that when I look at you, I see hope for the first time in three years."

The kiss was inevitable, soft, and tentative at first, then deeper as they both surrendered to the connection that had been building between them. Eva melted into Ben's embrace, her hands fisting in his shirt as he explored her mouth with gentle thoroughness.

When they finally broke apart, both were breathing hard. Eva could feel Ben's heart racing beneath her palms, could see the way his pupils had dilated with desire.

"We should probably get some rest," Ben said, though his voice lacked conviction.

"Probably," Eva agreed, but made no move to pull away from his embrace.

Instead, she began unbuttoning his shirt, her fingers working with scientific precision. "But first, I need to check the rest of your wounds."

Ben's breath caught as her hands spread across his chest, mapping the scars and muscle with careful attention. "Eva—"

"Shh," she whispered, pressing a kiss to a particularly nasty scar near his collarbone. "Let me take care of you."

Ben's control snapped. His hands tangled in her hair as he pulled her up for another kiss, this one hungrier, more desperate. Eva responded with equal fervor, her scientific reserve crumbling under the weight of desire and connection.

They moved together with an urgency born of shared trauma and mutual need. Eva's dress was elegant but impractical, and Ben's hands shook slightly as he worked the zipper, his touch reverent and careful.

"Are you sure?" he asked, his forehead pressed against hers.

"I've never been more sure of anything in my life," Eva replied, her voice steady despite the emotion threatening to overwhelm her.

What followed was tender and desperate, passionate and healing. They came together not as strangers thrown together by circumstance, but as two people who had found something precious in each other—understanding, acceptance, and the promise of redemption.

Afterward, they lay entwined on the couch, Eva's head on Ben's chest, listening to the steady rhythm of his heartbeat. The apartment's security systems hummed quietly around them, a reminder of the danger they faced, but for the moment, they were safe.

"Ben," Eva said quietly, "what happens now?"

"Now we stop the Phoenix Order," he replied, his fingers tracing patterns on her bare shoulder. "We use everything we know, everything we've learned, to prevent the deaths of four billion people."

"And after?"

Ben was quiet for a long moment. "I don't know," he admitted. "I've been so focused on running from my past that I never thought about having a future."

Eva lifted her head to look at him, seeing the vulnerability in his eyes. "I'd like to find out," she said softly. "If we survive this, I'd like to find out what happens next."

Ben's smile was the first genuine expression of happiness she'd seen from him. "I'd like that too."

As they settled back into each other's arms, Eva realized that something fundamental had changed between them. They were no longer just allies against a common enemy. They were partners in every sense of the word, bound together by trust, desire, and the shared determination to save the world from her father's madness.

The Phoenix Order had made a crucial mistake. They'd assumed that forcing Ben and Eva together would make them easier to capture, easier to control.

Instead, they'd created something far more dangerous than either individual could have been alone.

They'd created a partnership forged in fire, tempered by shared trauma, and strengthened by love.

And that partnership would be their undoing.

Outside, Berlin slept fitfully, unaware that its salvation might rest in the hands of two people who had found each other in the darkness.

The war was far from over, but Ben and Eva were no longer fighting it alone.

They were fighting it together.

And that made all the difference in the world.

# Chapter 13: The Data Analysis

The encrypted drive contained the blueprint for humanity's extinction.

Eva sat cross-legged on the floor of the safe house, surrounded by print-outs, laptops, and the detritus of twelve hours of intensive analysis. Ben paced behind her, his movements betraying the restless energy of a man confronting an impossible puzzle. The morning light filtering through the bulletproof windows illuminated the scope of their discovery: the Phoenix Order hadn't just created a bioweapon—they'd engineered the perfect apocalypse.

"Show me the protein structure again," Ben said, his voice tight with controlled tension.

Eva pulled up the holographic display on her laptop, the virus's molecular architecture rotating slowly in three-dimensional space. Even in the apartment's harsh lighting, the pathogen was beautiful in its complexity—a crystalline lattice of amino acids and genetic sequences that pulsed with malevolent purpose.

"This is my work," Eva said, her voice barely above a whisper. "The cellular penetration mechanism, the protein folding optimization—I designed every component of this delivery system."

Ben stopped pacing, placing his hands on her shoulders. "Eva, you couldn't have known—"

"Couldn't I?" Her laugh was bitter, self-recriminating. "Three years of research, and I never once questioned why we needed a delivery system this sophisticated. I was so focused on the elegance of the science that I ignored the implications."

The virus was a masterpiece of biological engineering. Eva's protein synthesis work had created a pathogen that could breach cellular defenses with unprecedented efficiency. The genetic targeting system allowed for population-specific lethality based on inherited markers. And the dispersal mechanism—airborne, stable, virtually undetectable—would ensure global distribution within hours.

"Walk me through the infection process," Ben said, his analytical mind demanding complete understanding of the threat they faced.

Eva's fingers traced the molecular structure on the screen. "Phase one is cellular infiltration. The virus uses my modified proteins to bypass the body's natural defenses. It's like giving the pathogen a key to every lock in the human immune system."

"How long from exposure to symptoms?"

"Seventy-two hours. The virus remains completely dormant during initial replication, giving it time to spread throughout the host's system before triggering the immune response."

Ben's face was grim. "Perfect for air travel. Infected passengers would carry the virus worldwide before anyone realized they were sick."

"That's just the beginning." Eva opened another file, her hands shaking as she scrolled through the technical specifications. "Phase two is where the real horror begins. The virus doesn't just kill cells—it reprograms them."

The screen filled with microscopic images that made Ben's stomach churn. Cellular structures twisted into alien configurations, DNA sequences rewritten according to the virus's genetic imperatives.

"Jesus Christ," Ben breathed. "It's not just destroying tissue—it's transforming it."

"Into what, I'm not entirely sure." Eva's scientific training warred with her revulsion. "The cellular regeneration patterns are unlike anything in nature. The virus forces rapid mutation, but the changes aren't random. They're directed, purposeful."

Ben studied the images, his intelligence training allowing him to process horrors that would have broken most people. "How does it kill?"

"That's the truly monstrous part." Eva's voice was clinical, detached—the only way she could discuss the weapon she'd unwittingly helped create. "The virus doesn't kill quickly. It causes progressive tissue breakdown while simultaneously triggering regeneration. The victim experiences cellular death and rebirth in an endless cycle."

"For how long?"

"Based on the projections, anywhere from seventy-two hours to two weeks. The suffering would be indescribable."

Ben closed his eyes, imagining billions of people enduring such agony. "And the genetic targeting?"

Eva opened another file, this one containing population demographic data that made the virus's true purpose horrifyingly clear. "The targeting sequences focus on specific genetic markers associated with geographic ancestry. Sub-Saharan African populations show a ninety-four percent mortality rate. South Asian populations are eighty-seven percent. Latin American populations, eighty-two percent."

"While European populations?"

"Twelve percent mortality rate. The virus is designed to spare what the Phoenix Order considers 'genetically superior' populations while eliminating everyone else."

Ben's hands clenched into fists. "Four billion people."

"Minimum." Eva's voice cracked slightly. "The virus is designed to evolve, to adapt to countermeasures. Each generation becomes more efficient, more lethal. Within six months, the mortality rates could be even higher."

The scope of the Phoenix Order's ambition was breathtaking in its cruelty. They weren't just planning mass murder—they were attempting to reshape human evolution itself, using science as a tool of racial purification.

"Show me the deployment schedule," Ben said.

Eva pulled up the operational timeline, and Ben felt his blood freeze. The Phoenix Order's plan was elegantly simple and virtually unstoppable. Twelve major airports, with simultaneous releases during peak holiday travel, and global distribution within twenty-four hours.

"Christmas Eve," Eva said unnecessarily. "Maximum passenger flow, minimum security awareness. The virus will be released through the ventilation systems at exactly 1200 GMT."

Ben studied the target list, his analytical mind calculating the implications. "JFK, Heathrow, Charles de Gaulle, Frankfurt, Dubai, Tokyo, LAX, São Paulo, Johannesburg, Mumbai, Beijing, Sydney. Every major continent, every significant population center."

"The dispersal pattern modeling shows global saturation within seventy-two hours," Eva continued. "By the time symptoms appear, the virus will have reached every major city on Earth."

"And the carriers?"

"Already in position." Eva's voice was hollow. "The Phoenix Order has been placing operatives for months. They're not suicide bombers—they're true believers who think they're saving humanity. They'll release the virus and return to their lives, never knowing they've triggered a global apocalypse."

Ben moved to the window, staring out at Berlin's bustling streets. Somewhere in the city, perhaps walking past at this very moment, was a Phoenix Order operative carrying a vial of death. The normalcy of the scene—people going to work, children playing, life continuing its ancient rhythm—made the coming horror almost incomprehensible.

"How do we stop it?" Ben asked, though he already suspected the answer.

"We can't." Eva's honesty was brutal. "The deployment is too distributed, too coordinated. Even if we could identify every operative, we'd need to prevent twelve simultaneous attacks across six continents. It's logistically impossible."

"Then we focus on the virus itself. Your mother's research—the counter-virus she was developing."

Eva pulled up another file, this one containing fragments of her mother's work. "She was brilliant, but she was working with incomplete information. The virus has evolved since her death. Her countermeasures might not be effective against the current iteration."

"But they're a starting point."

"Maybe." Eva's scientific skepticism warred with desperate hope. "The counter-virus would need to be as sophisticated as the original pathogen. We'd need to decode the genetic sequences, synthesize the appropriate proteins, and develop a delivery mechanism that could reach global populations faster than the Phoenix Order's weapon."

"How long would that take?"

"Under normal circumstances? Years. With unlimited resources and perfect conditions? Maybe months."

Ben's expression was grim. "We have sixty hours."

The impossibility of their situation settled over them like a suffocating blanket. The Phoenix Order had spent decades planning their attack, had unlimited resources, and had infiltrated governments and institutions worldwide. Ben and Eva were two people with a laptop, a few stolen files, and a safe house that might be discovered at any moment.

"There's something else," Eva said quietly. "The virus is designed to be self-modifying. Each generation adapts to environmental conditions and develops resistance to countermeasures. Even if we could create an effective counter-virus, the pathogen would evolve to defeat it."

"How quickly?"

"Based on the mutation rate projections, days. Maybe hours."

Ben sat heavily on the couch, the weight of their failure pressing down on him. "So we're talking about a weapon that can't be stopped, can't be countered, and will continue to evolve until it's killed everyone it's designed to kill."

"That's the assessment, yes."

The silence stretched between them, heavy with the implications of their discovery. Outside, Berlin continued its oblivious existence, unaware that extinction was approaching with clockwork precision.

"Ben," Eva said finally, "there might be another way."

"What?"

"The virus has a critical vulnerability. It's designed to be controllable, to respond to specific genetic commands. The Phoenix Order needs to be able to stop it if something goes wrong."

Ben's eyes sharpened. "A kill switch."

"Exactly. My mother's research mentions something called the 'Prometheus Protocol'—a genetic sequence that could theoretically neutralize the virus worldwide."

"Where is it?"

"Hidden." Eva's voice carried a note of desperate hope. "My mother was paranoid about Klaus discovering her work. She embedded the kill switch in something he'd never think to examine."

"What?"

Eva's smile was sad, nostalgic. "A lullaby. The song she used to sing to me when I was a child. She encoded the genetic sequence in the melody, hidden in plain sight."

Ben stared at her. "You're telling me that the key to saving four billion lives is hidden in a children's song?"

"My mother was brilliant, but she was also a parent. She knew that if something happened to her, I'd be the only one who could decode the message. She trusted me to understand."

"Can you remember it?"

Eva closed her eyes, and for a moment, Ben could see the frightened child she'd once been. "Every note. Every variation. She sang it to me thousands of times."

"Then we have a chance."

"A small one." Eva's scientific training demanded honesty. "Even if we can decode the kill switch, we'd need to synthesize it, weaponize it, and distribute it globally. We'd need laboratory facilities, manufacturing capabilities, and a delivery system that could reach every population center on Earth."

"One step at a time," Ben said. "First, we decode your mother's message. Then we figure out how to use it."

Eva nodded, but Ben could see the doubt in her eyes. The Phoenix Order had spent decades preparing for this moment. They had resources, connections, and the element of surprise. Ben and Eva had courage, desperation, and a children's song that might contain the key to humanity's survival.

It wasn't much. But it was all they had.

"Eva," Ben said softly, "your mother knew this day would come. She prepared for it, planned for it, and trusted you to finish what she started. She believed in you."

"I hope her faith wasn't misplaced."

"It wasn't." Ben's voice carried absolute conviction. "You're the most brilliant person I've ever met, and you're fighting for the right reasons. We're going to stop them."

Eva looked at the data spread around them—the virus specifications, the deployment schedules, the projected casualties. Four billion people, reduced to statistical abstractions by her father's madness.

"Ben," she said quietly, "promise me something."

"What?"

"If we fail, if the Phoenix Order succeeds—promise me you'll remember that we tried. That we didn't give up, even when it seemed impossible."

Ben moved to sit beside her, taking her hands in his. "We're not going to fail."

"You don't know that."

"Yes, I do." Ben's eyes met hers, and Eva saw something there that made her breath catch. "Because you're not fighting alone anymore. Because your mother's sacrifice won't be meaningless. Because sometimes, the right people in the right place at the right time can change everything."

Eva leaned into his embrace, drawing strength from his certainty. They were facing impossible odds, but they weren't facing them alone.

The Phoenix Order had made a critical mistake. They'd assumed that killing Eva's mother would end the threat to their plans.

Instead, they'd created something far more dangerous.

They'd created a daughter with nothing left to lose, a partner willing to risk everything, and a mother's love that transcended death itself.

The lullaby would save the world.

Or they would all die trying.

But they would not go quietly into the darkness.

They would fight.

And sometimes, that was enough.

# Chapter 14: The Network

B en's fingers hovered over the encrypted keyboard, each keystroke potentially his last. The communication protocols he'd memorized during his CIA years felt alien now, like a language he'd once spoken fluently but had forgotten through disuse. Three years of academic exile had dulled his edge, but the muscle memory of tradecraft was returning with terrifying clarity.

"Are you sure about this?" Eva asked, her voice tight with concern. She stood behind him, her hands resting on his shoulders, providing anchor and comfort in equal measure. "If the Phoenix Order has infiltrated the intelligence community..."

"Then we need to know how deep the corruption goes," Ben replied, his voice steady despite the knot of dread in his stomach. "I can't fight an enemy I don't understand."

The safe house's secure communication system was military-grade, designed to defeat even nation-state surveillance. But Ben knew that the Phoenix Order's resources might exceed those of most governments. Every transmission was a calculated risk, every contact a potential death sentence.

He began with Marcus Webb, a CDC epidemiologist who had collaborated with the CIA on bioweapons assessments. Webb was solid, reliable, the kind of bureaucratic lifer who'd never betray his country for ideology or money. If anyone could verify the Lethe Virus's capabilities, it would be him.

The encrypted message was brief, coded in language that would seem innocuous to casual observers:

Marc - Academic research project requires consultation on viral pathology. Protein synthesis, cellular targeting, and airborne dispersal mechanisms. Urgent timeline. Can you assist? - B

Ben hit send and immediately felt the weight of exposure. The message was bouncing through multiple servers, encrypted and re-encrypted. Still, somewhere in the digital darkness, the Phoenix Order might be watching.

"Now we wait," he said, leaning back in his chair.

Eva's hands tightened on his shoulders. "How long?"

"Marcus always responds within an hour. He's obsessively punctual." Ben glanced at the clock: 14:23. "If we don't hear from him by 15:30, we'll know he's compromised."

The minutes crawled by with agonizing slowness. Eva busied herself with her mother's research, trying to decode the lullaby's hidden genetic sequences. Ben watched the communication terminal, his trained eye cataloging every system alert, every network fluctuation.

At 14:47, the terminal chimed.

Ben - Interesting timing. CDC received unusual queries about viral countermeasures last week. Official channels, but wrong clearance protocols. Someone's asking questions they shouldn't. Need to talk. Secure channel 7, 1600 hours. - M

Ben's blood chilled. "Someone's been asking the CDC about viral countermeasures."

"The Phoenix Order?"

"Or someone trying to stop them." Ben's analytical mind processed the implications. "The wrong clearance protocols suggest either incompetence or deliberate deception."

Eva looked up from her work. "Could be allies. Other people who've discovered the plot."

"Or it could be the Phoenix Order testing CDC responses." Ben's paranoia was now fully engaged. "In intelligence work, coincidences are usually enemy action."

The secure channel connection was established at precisely 16:00. Marcus Webb's face appeared on screen, his usual bureaucratic composure cracked by visible stress.

"Ben, where the hell are you?" Marcus's voice carried three years of accumulated worry. "After you disappeared from the university, federal agents started asking questions. They showed me your photo, wanted to know about any communications we'd had."

"What did you tell them?"

"The truth. That we'd collaborated on bioweapons assessments, that you'd been out of contact for three years." Marcus leaned closer to the camera. "Ben, they knew things about our work that aren't in any official files. Classified details that only a handful of people should have access to."

Ben's hands clenched into fists. "What kind of details?"

"The Damascus operation. The civilian casualties. Your psychological evaluation after the incident." Marcus's voice dropped to a whisper. "They knew about your guilt complex, your tendency toward self-isolation. They were profiling you, Ben. Building a psychological dossier."

"Jesus." Ben's worst fears were being confirmed. "Marcus, I need you to run a theoretical scenario. Airborne viral pathogen, cellular targeting based on genetic markers, global deployment through airport ventilation systems."

Marcus's face went pale. "That's not theoretical, is it?"

"Just run the scenario."

"Jesus Christ, Ben. The mortality rates would be—"

"I know what the mortality rates would be." Ben's voice was sharp with urgency. "I need to know about countermeasures. Is there any way to stop something like that once it's been released?"

Marcus was quiet for a long moment, his epidemiologist's mind working through the implications. "Not through conventional means. The distributed nature of the attack, the genetic targeting, the airborne dispersal—it's designed to be unstoppable."

"What about a counter-virus?"

"Theoretically possible, but the logistics would be nightmarish. You'd need to synthesize the countermeasure, weaponize it, and distribute it globally faster than the original pathogen could spread. And that's assuming you could develop something effective against a constantly evolving target."

Ben felt Eva's hand squeeze his shoulder. "Marcus, if someone were planning such an attack, who would they need to compromise?"

"Everyone." Marcus's honesty was brutal. "CDC, WHO, pharmaceutical companies, government health agencies, airport authorities, intelligence services. The scope of infiltration would have to be massive."

"And if they had that level of infiltration?"

"Then we'd be fucked." Marcus's professional composure cracked entirely. "Ben, please tell me this is just a theoretical exercise."

Before Ben could answer, the communication link went dead. Not a typical disconnection, but a sudden, violent termination that suggested outside interference.

"Marc?" Ben tried to reestablish the connection, but the secure channel was completely dark. "Shit."

Eva's face was pale. "What happened?"

"Someone cut the connection. Either Marcus is compromised, or—"

The safe house's perimeter alarms began shrieking. Ben's blood froze as he saw the surveillance display: black SUVs surrounding the building, tactical teams deploying with military precision.

"They traced the communication," Eva said, her voice hollow with realization.

"Move. Now." Ben was already grabbing their emergency supplies, his training taking over completely. "Everything we can carry. Leave the rest."

They had maybe three minutes before the tactical teams breached the building. Ben's mind raced through escape routes, contingency plans, and the careful preparations he'd made for exactly this scenario.

"The roof," he said, leading Eva toward the hidden ladder. "There's a zip line to the building across the street."

As they climbed, Ben could hear the tactical teams entering the building below. Professional voices, coordinated movement, and the sound of doors being kicked in with systematic efficiency.

The roof was windswept and exposed, but Ben had prepared for this. The zip line was barely visible, a thin cable stretching across the street to the fire escape of another building.

"I can't," Eva said, staring at the gap. "I'm not trained for this."

"You are now." Ben clipped the harness around her waist, his hands steady despite the chaos below. "Trust me."

Eva nodded, and Ben saw the same courage that had driven her to steal her father's files, to risk everything for humanity's survival. She stepped off the roof without hesitation, sliding across the cable with grace despite her terror.

Ben followed, the zip line singing under his weight. Behind him, the tactical teams reached the roof just as he landed on the opposite building.

"There!" A voice shouted. "Northwest building, fire escape!"

Ben and Eva ran, their footsteps echoing off the metal stairs. They reached the street level just as the first Phoenix Order vehicles rounded the corner.

"This way," Ben said, pulling Eva into the crowd of afternoon shoppers. "We blend in, move with the crowd, get to the secondary safe house."

But as they walked, Ben's trained eye caught details that made his blood run cold. The man reading a newspaper at the café—the same one who'd been watching his apartment. The woman walking her dog was too alert, too focused on their movement. The delivery truck was parked with perfect sight lines—engine running, driver missing.

"They're everywhere," Eva whispered, reading his expression.

"Not everywhere," Ben replied, but his voice lacked conviction. "Just... more places than I expected."

They were walking through a surveillance network that extended far beyond what Ben had imagined. The Phoenix Order hadn't just infiltrated the intelligence community—they'd created a parallel surveillance state, watching, waiting, and coordinating with resources that rivaled those of major governments.

Ben's phone buzzed with an incoming message from an encrypted number:

Dr. Carter - Your CDC contact experienced an unfortunate accident. Cardiac arrest is very sudden. His research files are being reviewed for security violations. We hope you understand the importance of discretion. - Phoenix

Ben's hands shook as he showed Eva the message. Marcus Webb was dead, eliminated because he'd verified their fears about the Lethe Virus's capabilities.

"They killed him," Eva said, her voice barely audible. "Just for talking to us."

"They're sending a message," Ben replied. "Anyone who helps us dies."

As they walked through Berlin's streets, Ben felt the weight of isolation settling on his shoulders. The intelligence network he'd trusted, the colleagues he'd worked with, the institutions he'd served—all of it was compromised or eliminated.

His phone buzzed again. Another message:

Ben - Sarah Chen's funeral is tomorrow. Thought you should know. Her research files were very interesting. Particularly her notes about the Phoenix Order infiltration. She was closer to the truth than we realized. Too close. - Director Harrison

Ben stared at the message, his mind reeling. Director Harrison was his former CIA supervisor, the man who had recruited him, trained him, and saved him from the fallout in Damascus. If Harrison was compromised...

"We're completely alone," Ben said quietly.

"Not completely." Eva's hand found his, their fingers intertwining. "We have each other."

Ben nodded, but the reality of their situation was crushing. They were facing a global conspiracy with unlimited resources, government connections, and the ability to eliminate anyone who threatened their plans. They had courage, desperation, and a children's song that might contain the key to humanity's salvation.

The odds were impossible. The enemy was everywhere. The stakes couldn't be higher.

But as Ben looked at Eva's determined face, he realized that sometimes impossible odds were exactly what heroes were made for.

"Eva," he said quietly, "your mother encoded that lullaby for a reason. She knew this day would come, knew you'd need the kill switch. She trusted you to finish what she started."

"I just hope I'm strong enough."

"You are." Ben's voice carried absolute conviction. "You're the strongest person I've ever met. And you're not fighting alone."

They walked through Berlin's afternoon crowds, two people against the world, carrying the weight of four billion lives on their shoulders. The Phoenix Order had made their message clear: anyone who helped them would die.

But Ben and Eva had a message of their own: they would not go quietly into the darkness.

They would fight.

And sometimes, that was enough.

The network was compromised. The allies were dead. The enemy was every-where.

But love, courage, and a mother's sacrifice could move mountains.

And sometimes, mountains needed to be moved.

The war was far from over.

But it wasn't over yet.

And as long as they were breathing, as long as they had each other, there was hope.

The Phoenix Order had made a critical mistake.

They'd assumed that killing their allies would leave them helpless.

Instead, they'd made the fight personal.

And that was the most dangerous mistake of all.

# Chapter 15: The Prague Gambit

The train from Berlin to Prague should have been a six-hour journey of anonymous travel through the European countryside. Instead, it became a masterclass in paranoia as Ben and Eva watched every passenger, catalogued every stop, and analyzed every face for signs of Phoenix Order surveillance.

"Third car, window seat, blue jacket," Eva whispered, her voice barely audible above the train's rhythmic clatter. "He's been watching us since Dresden."

Ben glanced casually toward the indicated passenger, his trained eye confirming Eva's assessment. The man was good, professional in his posture, natural in his movements, the kind of low-profile surveillance that suggested extensive training. But Eva's scientific mind had spotted the pattern: too-consistent newspaper reading, periodic glances timed to their movements, the subtle repositioning that maintained optimal sight lines.

"You're getting better at this," Ben said, impressed despite their circumstances.

"I'm a quick learner." Eva's smile was grim. "Survival is excellent motivation."

The irony wasn't lost on either of them. Eva, the sheltered scientist, was developing the same paranoid awareness that had kept Ben alive through years of intelligence work. The Phoenix Order's relentless pursuit was forging her into something she'd never imagined becoming—a skilled operative capable of matching wits with professional killers.

"David Roth is different," Ben said, returning to their whispered conversation about the Prague meeting. "Mossad trains its people differently from the CIA. He's survived situations that would have killed most operatives."

"You trust him?"

"With my life. He pulled me out of Damascus when everything went to hell. If anyone can verify the Phoenix Order's capabilities and help us find allies, it's David."

Eva nodded, but Ben could see the doubt in her eyes. After Marcus Webb's murder, after the systematic elimination of every potential ally, trust had become a luxury they couldn't afford.

The train pulled into Prague's central station at 19:47, seventeen minutes behind schedule. Ben's operational instincts were screaming warnings as they disembarked into the evening crowd—too many people positioned too perfectly, too many eyes that seemed to track their movement.

"We're expected," Eva said quietly, her newly developed awareness picking up the same signals.

"Maybe. Or maybe Prague is just naturally paranoid." Ben guided her through the station's Gothic architecture, using the crowd as cover while scanning for immediate threats. "David chose the meeting location. If it's compromised, he'll know."

The Charles Bridge at midnight was Prague's most romantic tourist destination, but Ben had chosen it for tactical reasons. The medieval stone bridge provided multiple escape routes, clear sight lines, and sufficient late-night foot traffic to offer cover without compromising security.

David Roth was waiting at the center of the bridge, his silhouette unmistakable against the baroque statues that lined the walkway. Even in civilian clothes, the Israeli operative carried himself with the coiled alertness of a professional killer.

"Ben." David's English carried the slight accent of someone who'd learned the language in war zones. "You look like shit."

"Retirement doesn't suit me." Ben embraced his former colleague, genuinely glad to see a friendly face. "David, this is Dr. Eva Richter. She's the one with the intelligence about the Phoenix Order."

David's assessment of Eva was swift and thorough, cataloguing her appearance, body language, and potential threat level in seconds. "Doctor, Ben tells me you have information about a bioweapons conspiracy."

"Four billion people are going to die in thirty-six hours," Eva said simply. "Unless we can stop them."

David's expression didn't change, but Ben caught the slight tension in his shoulders. "Show me."

Eva opened her laptop, displaying the viral specifications and deployment schedules. David studied the data with the focused intensity of someone who'd spent years evaluating existential threats.

"This is authentic," he said finally. "The technical sophistication, the operational planning—this isn't the work of amateurs."

"Can Mossad help us?" Ben asked.

David's pause was answer enough. "Ben, there's something you need to know. The Phoenix Order has been on our radar for months. We've been tracking their financial networks, their recruitment patterns, their government connections."

"And?"

"They're not just another terrorist organization. They're a state-level actor with resources that rival major intelligence agencies. Our attempts to penetrate their operations have been... unsuccessful."

Ben felt the familiar weight of disappointment. Another ally, another dead end. "So you can't help us."

"I didn't say that." David's smile was predatory. "Mossad has been developing countermeasures for bioweapon attacks. If your virus specifications are accurate, we might have something useful."

Eva leaned forward, her scientific interest overriding her wariness. "What kind of countermeasures?"

"Classified," David replied, but his expression suggested hope. "But if you're willing to share your research, we might be able to develop something effective."

The negotiation was interrupted by the distinctive sound of motorcycle engines—multiple bikes, approaching from both ends of the bridge with coordinated precision.

"Shit," David muttered, his hand moving instinctively toward his concealed weapon. "We need to move. Now."

But it was too late. The motorcycles had positioned themselves to block the bridge's exits, their riders dismounting with military efficiency. Ben counted six operatives, all armed, all moving with the coordinated precision of a professional hit squad.

"David," Ben said quietly, "please tell me you have an exit strategy."

"I did." David's voice was grim. "But it looks like the Phoenix Order anticipated our meeting."

The lead operative approached with casual confidence, his weapon visible but not overtly threatening. "Dr. Roth, Dr. Carter, Dr. Richter. You're all under arrest for crimes against international security."

"What crimes?" Eva demanded, her scientific mind rejecting the illogical accusation.

"Bioterrorism, conspiracy to commit mass murder, theft of classified materials." The operative's smile was coldly professional. "The Phoenix Order has provided compelling evidence of your activities."

Ben's blood chilled. They weren't just being hunted—they were being framed. The Phoenix Order was positioning them as the terrorists, the masterminds behind the very plot they were trying to stop.

"You're making a mistake," Ben said, his voice steady despite the rage building in his chest.

"No mistakes," the operative replied. "Just justice."

David moved first, his Mossad training kicking in with lethal precision. His weapon appeared in his hand like magic, and the lead operative dropped with a bullet between his eyes.

"Run!" David shouted, providing covering fire as the other operatives scrambled for position.

Ben grabbed Eva's hand and sprinted toward the bridge's Gothic towers, using the medieval stonework as cover. Behind them, gunfire erupted in the night air, bullets striking ancient stone and sending fragments flying.

"This way!" Eva pulled Ben toward a maintenance ladder that led to the upper levels of the bridge. "I studied Prague's architecture in graduate school. There's a route through the bell tower."

Ben followed, marveling at Eva's quick thinking under pressure. The woman who'd been a sheltered scientist days ago was now thinking like a tactical operative, using her knowledge to find escape routes that his military training had missed.

The bell tower was cramped and dark, but it connected to the rooftops of Prague's Old Town. Eva moved through the medieval passages with surprising confidence, her academic knowledge transformed into practical survival skills.

"There," she pointed to a narrow gap between buildings. "We can cross to the next roof."

The jump was treacherous, a six-foot gap across a three-story drop. Ben went first, his body remembering the parkour training from his CIA days. Eva followed without hesitation, her leap perfect despite her obvious fear.

They landed on the adjacent roof just as the first Phoenix Order operative reached the bell tower behind them. The pursuit had become a chase across Prague's ancient skyline, with medieval architecture providing both obstacles and opportunities.

"Keep moving," Ben said, guiding Eva across the slanted rooftops. "They'll have the streets covered."

The chase became a nightmare of leaping between buildings, sliding down drainage pipes, and using architectural features as handholds. Eva's hidden athleticism impressed Ben—she moved with the grace of someone who had been athletic in her youth, her body recalling skills that academic life had temporarily buried.

"There!" Eva pointed to a gap in the pursuit line. "We can reach the river."

They slid down a steep roof, using the ancient tiles as an improvised slide. The Phoenix Order operatives were skilled, but Prague's medieval architecture favored defenders who were familiar with the terrain.

The Vltava River stretched below them, dark and cold in the December night. A tourist boat was moored at the dock, its engines running, crew preparing for departure.

"Can you swim?" Ben asked.

"About to find out," Eva replied, and jumped.

The water was shockingly cold, but they both surfaced quickly. The tourist boat's crew, startled by their dramatic arrival, helped them aboard without question.

"American tourists," Ben gasped, his cover story automatic. "Robbed by a motorcycle gang. Need to get to the embassy."

The boat captain, a grizzled Czech who'd probably seen stranger things, nodded and gunned the engines. As they pulled away from the dock, Ben could see the Phoenix Order operatives reaching the waterfront, their frustrated gestures visible even in the darkness.

"We made it," Eva said, her voice filled with exhausted relief.

"Did we?" Ben's expression was grim. "David's dead, our equipment is compromised, and we're no closer to stopping the Phoenix Order."

Eva moved closer to him, her wet clothes clinging to her body, her hair streaming water. "Ben, we're alive. We escaped. We're still fighting."

"For now." Ben's voice carried the weight of their impossible situation. "But we've lost everything. No allies, no resources, no way to synthesize the counter-virus even if we can decode your mother's lullaby."

Eva's hands framed his face, her touch warm despite the cold water. "We have each other."

"Is that enough?"

"It has to be." Eva's voice was fierce with determination. "Because the alternative is four billion deaths."

Ben looked into her eyes and saw something that made his chest tighten. Not just courage, but hope. Despite everything they'd lost, despite the impossible odds, Eva still believed they could succeed.

"You're incredible," he said softly. "You know that?"

"I'm terrified," Eva admitted. "But I'm not giving up. My mother died for this. David died for this. We're not letting their sacrifices be meaningless."

The moment stretched between them, electric with tension that had nothing to do with their recent escape. They were alone on a tourist boat in the middle of the Prague river, soaked and exhausted, hunted by a global conspiracy. But they were also alive, together, and still fighting.

"Eva," Ben said, his voice rough with emotion.

"I know," she whispered.

The kiss was inevitable, desperate, and perfect. They came together with the passion of people who'd faced death and chosen life, who'd found each other in the darkness and refused to let go.

When they broke apart, both were breathing hard. Eva's eyes were bright with tears and determination.

"We're going to stop them," she said. "Somehow, we're going to decode my mother's lullaby and save the world."

"Together," Ben agreed, his voice steady for the first time in hours.

"Together," Eva confirmed.

As the tourist boat carried them through Prague's ancient waterways, Ben realized that the Phoenix Order had made a crucial miscalculation. They'd assumed that eliminating allies would leave him and Eva helpless.

Instead, they'd forged the survivors into something stronger than either could have been alone.

The war was far from over. But Ben and Eva were no longer just fighting for humanity's survival.

They were fighting for each other.

And that made them more dangerous than any weapon the Phoenix Order could deploy.

The Prague gambit had failed to capture them.

But it had succeeded in creating something the Phoenix Order should fear far more than two isolated academics.

It had created two people with nothing left to lose except each other.

And people like that could move mountains.

The Phoenix Order was about to learn that love was the most dangerous weapon of all.

The hunt continued. But now, the hunters were becoming the hunted.

And in thirty-four hours, the world would learn which side had made the better choices.

The chase across Prague's rooftops was over.
But the real battle was just beginning.

# Chapter 16: The Father's Hunt

D r. Klaus Richter stood in the Phoenix Order's operational command center, studying the wall of monitors that displayed real-time intelligence from across Europe. Each screen showed a different piece of the puzzle: surveillance footage from Prague, intercepted communications, predictive tracking algorithms, and the systematic elimination of every potential ally his daughter might seek.

But Klaus wasn't looking at the technology. He was looking at a childhood photograph of Eva, age seven, building a molecular model in his laboratory. Even then, she'd possessed the brilliant mind that had made her invaluable to the Phoenix Order's plans. And even then, she'd shown the stubborn moral courage that now made her the greatest threat to humanity's evolution.

"The Prague operation was a failure," Wilhelm von Hess said, his voice carrying the cold displeasure of a man unaccustomed to setbacks. "Three operatives dead, our asset compromised, and your daughter still running free with classified intelligence."

Klaus didn't respond immediately. His eyes remained fixed on the photograph, as he remembered the day it was taken. Eva had been so proud of her first successful protein synthesis, so eager to share her discovery with the father she adored. That same scientific brilliance now threatened to unravel everything they'd worked toward.

"Klaus," von Hess continued, his tone sharpening. "I'm beginning to question whether you're the appropriate choice to lead this hunt."

"I know my daughter better than anyone," Klaus replied, his voice steady despite the turmoil in his chest. "Her behavioral patterns, her psychological vulnerabilities, her emotional attachments. She's predictable in ways that you and your operatives don't understand."

"Predictable?" Von Hess's laugh was bitter. "She's evaded our best hunters for four days. She's eliminated our assets, compromised our security, and turned your own research against us. That doesn't sound predictable to me."

Klaus finally turned from the photograph, his pale eyes meeting von Hess's with the intensity that had made him legendary in scientific circles. "Eva is responding exactly as her psychological profile suggests. She's seeking allies, trying to build a network of resistance, believing that traditional institutions can be trusted to help her."

"And that makes her predictable?"

"It makes her vulnerable." Klaus moved to the central display, his fingers calling up a psychological assessment file. "Eva's personality matrix shows classic patterns of idealistic authority dependence. She trusts institutions, believes in official channels, and seeks validation from established experts. It's a weakness that can be exploited."

The file contained twenty-eight years of psychological observations, compiled by a father who had studied his daughter with the same analytical precision he applied to his viral research. Every behavioral pattern, every emotional trigger, every psychological vulnerability was catalogued and cross-referenced.

"She'll seek out academic contacts," Klaus continued, his voice taking on the clinical detachment of a researcher discussing a specimen. "Scientists who might validate her concerns, intelligence operatives who might provide resources, and government officials who might take action. Each contact point is an opportunity for interception."

Von Hess studied the psychological profile with interest. "You're suggesting we let her run, let her expose herself through her own attempts to find help?"

"I'm suggesting we use her predictability against her." Klaus's voice carried a note of sadness that he quickly suppressed. "Eva is brilliant, but she's also naive. She believes in the fundamental goodness of people, in the power of truth to overcome deception. These beliefs make her strong, but they also make her exploitable."

The command center's communication system chimed with an incoming intelligence report. Klaus accepted the transmission, his expression darkening as he read.

"Vienna," he said quietly. "She's heading for Vienna."

"How can you be certain?"

Klaus pulled up a map of Central Europe, his finger tracing a route through the continent. "After Prague, she'll avoid major cities where our surveillance is strongest. But she'll need resources, allies, and scientific expertise. Vienna is home to the International Atomic Energy Agency, the United Nations Office, and multiple academic institutions. It's exactly where her psychological profile suggests she would go."

Von Hess nodded slowly. "And we'll be waiting."

"We'll be hunting." Klaus's voice hardened. "Eva may be my daughter, but she's also a threat to human evolution. Personal feelings cannot be allowed to compromise the mission."

But even as he spoke the words, Klaus felt the weight of his own deception. Personal feelings were exactly what was compromising the mission. Every operational decision was filtered through twenty-eight years of paternal love, every strategic choice influenced by memories of bedtime stories and childhood laughter.

The truth was that Klaus Richter, the brilliant scientist and devoted Phoenix Order operative, was being destroyed by the need to hunt his own daughter.

"Sir," a technician called out, "we're receiving reports from our Vienna assets. Multiple government agencies are conducting unusual security sweeps. Someone's been asking questions about bioweapons countermeasures."

Klaus's chest tightened. "Show me."

The screens are filled with surveillance footage from Vienna's International District. Intelligence operatives move with coordinated precision, while scientific facilities undergo unexpected security reviews, and a systematic investigation is conducted into every institution that might provide aid to a desperate biochemist.

"She's not just seeking help," Klaus realized. "She's already found it."

Von Hess's expression was grim. "Then we need to move immediately. Every hour of delay increases the risk of exposure."

"No." Klaus's voice was firm. "If we move too quickly, we'll drive her underground. Eva's survival instincts are stronger than you realize. She'll disappear into the European underground, and we'll never find her."

"Then what do you suggest?"

Klaus turned back to the psychological profile, his analytical mind working through the implications. "We use her greatest strength against her. Eva's moral courage, her determination to save lives, her willingness to sacrifice herself for the greater good. These qualities make her a hero, but they also make her predictable."

"Explain."

"She won't run indefinitely. Eva believes she has a moral obligation to stop the Phoenix Order, to prevent the deaths of four billion people. That obligation will drive her to take risks, to expose herself, to choose confrontation over safety."

Klaus pulled up another file, this one containing detailed intelligence about Benjamin Carter. "And she's not alone. Carter's psychological profile shows similar patterns. Guilt-driven heroism, compulsive need for redemption, and willingness to sacrifice personal safety for perceived duty. Together, they're twice as dangerous but also twice as predictable."

"You're suggesting we let them come to us?"

"I'm suggesting we give them no choice." Klaus's voice carried the cold certainty of a man who'd studied his targets for decades. "Eva needs laboratory facilities to decode her mother's research. Carter needs intelligence resources to understand our operational structure. There are only a limited number of institutions in Europe that can provide both."

Von Hess's smile was predatory. "And we control most of them."

"We control all of them." Klaus's honesty was brutal. "Every major research facility, every government agency, every academic institution that might help them. They're not just being hunted—they're being herded toward a confrontation they can't avoid."

The command center fell silent as the implications sank in. The Phoenix Order hadn't just been pursuing Eva and Ben—they'd been manipulating them, using their heroic instincts to drive them toward a trap that would eliminate both threats simultaneously.

"There's something else," Klaus continued, his voice dropping to barely above a whisper. "Eva's mother encoded the counter-virus formula in a lullaby. If Eva can decode it, if she can synthesize the antidote, she might actually be able to stop the Lethe Virus."

Von Hess's face went pale. "You're telling me that your daughter has the ability to destroy everything we've worked for?"

"I'm telling you that Ingrid was more clever than we realized." Klaus's voice carried a note of grudging admiration. "She didn't just develop countermeasures—she ensured that only Eva could access them. The genetic code is embedded in childhood memories, in emotional associations that can't be extracted through torture or coercion."

"Then we need to kill her immediately."

"No." Klaus's voice was sharp with protective instinct, then softer as he regained control. "Killing Eva would be a mistake. She's the only person who can decode the lullaby, but she's also the only person who can verify its effectiveness. We need her alive long enough to ensure the counter-virus is real, then eliminate her before she can use it."

Von Hess studied Klaus with the calculating gaze of a predator evaluating another predator. "You're playing a dangerous game, Klaus. Your emotional attachment to your daughter is compromising your judgment."

"My emotional attachment is the key to stopping her." Klaus's voice carried the weight of absolute certainty. "I know Eva better than she knows herself. I know her fears, her hopes, her psychological triggers. I know exactly how to manipulate her into making the choices we need her to make."

"And Carter?"

"Carter is already broken. The Damascus guilt, the isolation, the constant fear—he's operating on pure instinct. Eva is his anchor, his reason for continuing. If we can separate them, manipulate their emotional bond, we can control both."

Klaus turned back to the monitors, his reflection ghostlike in the dark screens. "The hunt isn't just about elimination. It's about psychological warfare. We're not just hunting terrorists—we're hunting my daughter and the man she's falling in love with."

"Love?" Von Hess's voice carried a note of disgust. "You're suggesting we use emotional manipulation?"

"I'm suggesting we use every weapon at our disposal." Klaus's voice was cold with scientific detachment. "Love makes people vulnerable, desperate, willing to sacrifice everything for each other. It's the most powerful force in human psychology, and the most easily exploited."

The command center's communication system chimed again. Klaus accepted the transmission, his expression showing the first genuine emotion he'd displayed—satisfaction.

"Vienna confirmed," he said. "They've made contact with a WHO researcher. Dr. Elisabeth Müller, biochemical weapons specialist. She's agreed to provide laboratory access."

Von Hess's smile was predatory. "And Dr. Müller's loyalties?"

"Purchased three months ago." Klaus's voice carried the satisfaction of a chess master whose trap was finally springing. "She's Phoenix Order, instructed to gain their trust and then eliminate them."

"Perfect."

"Almost perfect." Klaus's voice carried a note of caution. "Eva is brilliant, and she's becoming more paranoid. She might suspect a trap."

"Then we make sure she doesn't have time to think." Von Hess moved to the tactical display, his fingers calling up deployment schedules. "How long do we have?"

"Thirty-two hours until release." Klaus's voice was steady despite the magnitude of what he was discussing. "If Eva can decode the lullaby and synthesize the counter-virus, she'll need at least forty-eight hours. The timeline favors us."

"Unless she's already started the process."

Klaus shook his head. "She'll need specialized equipment, controlled conditions, raw materials that can only be obtained through official channels. The counter-virus isn't something that can be created in a basement laboratory."

"Then we have her."

"We have her." Klaus's voice carried the weight of finality. "Eva Richter, my brilliant daughter, is about to walk into a trap that will destroy her and save humanity's future."

But even as he spoke the words, Klaus felt something breaking inside his chest. The rational scientist understood the necessity of Eva's elimination. The Phoenix Order operative accepted the need to sacrifice family for the greater good. But the father—the man who'd sung lullabies and bandaged scraped knees and watched his daughter take her first steps—was dying with every decision.

"Klaus," von Hess said quietly, "I need to know that you can do this. That when the moment comes, you'll be able to eliminate your own daughter."

Klaus looked at the childhood photograph one last time, then turned it face down on the desk. "Eva ceased to be my daughter the moment she chose to oppose human evolution. She's just another enemy now."

The lie came easily, practiced, and smooth. But Klaus knew the truth. Eva would always be his daughter, no matter what choices she made. And when the moment came to eliminate her, it would destroy what remained of his soul.

But that was the price of evolution. That was the sacrifice required to create a better world.

The Phoenix Order demanded everything. Family, love, conscience, and humanity itself.

And Klaus Richter, the man who'd once been a father, was prepared to pay that price.

The hunt was entering its final phase.

And only one side would survive.

The father in Klaus hoped it would be Eva.

The monster he'd become knew it couldn't be.

The war between love and ideology was about to reach its climax.

And in thirty-two hours, the world would learn which force was stronger.

The hunt continued.

But now, the hunter was hunting his own heart.

And that was the most dangerous prey of all.

# Chapter 17: The Mother's Lullaby

The Vienna safe house was a cramped studio apartment in the city's Second District, hidden above a bakery that filled the building with the comforting scent of fresh bread. But Eva barely noticed the normalcy surrounding her as she hunched over her laptop, surrounded by printouts and notebooks, her mother's lullaby playing on endless repeat through noise-canceling headphones.

Phoenix rising from the ashes, born of fire and ancient light...

The melody was hauntingly beautiful, a complex composition that had seemed like nothing more than a childhood comfort song. But Eva's trained ear was beginning to detect patterns beneath the surface—mathematical relationships. These harmonic structures corresponded to genetic sequences.

"Talk to me," Ben said, setting a cup of coffee beside her. His voice carried the controlled tension of someone watching a genius work on humanity's salvation while knowing that time was running out.

Eva removed the headphones, her eyes bright with exhaustion and discovery. "It's all here, Ben. Every note, every variation, every melodic phrase—it's a complete genetic blueprint."

"For what?"

"A counter-virus." Eva's voice trembled with the magnitude of what she was uncovering. "My mother didn't just hide a formula in the lullaby. She encoded an entire viral genome, protein by protein, sequence by sequence."

Ben leaned closer, studying the complex notations Eva had been making. Musical notes were transformed into chemical formulas, harmonic progressions became genetic sequences, and the simple childhood song revealed itself as the most sophisticated piece of biochemical engineering ever conceived.

"How is that possible?"

"Music and genetics follow similar mathematical principles," Eva explained, her scientific enthusiasm overriding her exhaustion. "Harmonic relationships, pattern recognition, structural complexity. My mother was a musical prodigy before she became a scientist. She understood both languages fluently."

Eva pulled up a molecular visualization program, her fingers dancing across the keyboard as she translated musical notation into genetic code. The screen filled with a three-dimensional protein structure that seemed to pulse with inner light.

"This is it," she breathed. "The counter-virus. Designed to target the Lethe Virus at the molecular level."

Ben studied the rotating structure, his intelligence training allowing him to appreciate the elegant complexity of what he was seeing. "How does it work?"

"It's brilliant." Eva's voice carried genuine admiration for her mother's genius. "The Lethe Virus uses my protein synthesis work to penetrate cellular defenses. But my mother's counter-virus uses the same pathways to deliver a genetic payload that neutralizes the original pathogen."

"Like a key that fits the same lock?"

"Exactly. But instead of opening the door for destruction, it opens it for healing." Eva's fingers traced the molecular structure on the screen. "The counter-virus doesn't just neutralize the Lethe Virus—it repairs the cellular damage it causes."

Ben felt hope stirring in his chest for the first time in days. "So we can stop it. We can actually stop the Phoenix Order."

"Maybe." Eva's scientific training demanded honesty. "The counter-virus is theoretically perfect, but theory and practice are different things. I'd need to synthesize it, test it, and verify its effectiveness against the current iteration of the Lethe Virus."

"What do you need?"

Eva's expression darkened. "Everything I don't have. A Level 4 biosafety lab-oratory, industrial-grade synthesis equipment, samples of the original virus for testing, and computational resources that can model molecular interactions in real-time."

"And time?"

"Forty-eight hours minimum, assuming perfect conditions and no complica-tions." Eva's voice carried the weight of impossibility. "Ben, we have thirty-one hours before the Phoenix Order releases the virus. Even if we could access the necessary facilities, the timeline is—"

"Impossible," Ben finished. "But not necessarily insurmountable."

Eva looked at him with curiosity and growing hope. "What are you thinking?"

"I'm thinking that there's only one facility in Europe that has everything you need." Ben's voice was grim with the implications. "The Phoenix Order's main laboratory in the Bavarian Alps."

"My father's laboratory." Eva's face went pale. "Ben, that's suicide. The security is military-grade, the location is isolated, and Klaus will be expecting us."

"Maybe. But it's also the only place where we can synthesize the counter-virus and deploy it before the Christmas Eve deadline."

Eva stood up, pacing the small apartment as her mind processed the implica-tions. "Even if we could infiltrate the facility, even if we could access the laboratory equipment, we'd need samples of the original virus. And those are stored in the most secure section of the complex."

"Walk me through the layout."

Eva pulled up architectural blueprints on her laptop, her intimate knowledge of the facility evident in every detail. "The monastery is built into the mountain-side, with multiple levels extending deep into the rock. The main laboratory is on Sub-Level 3, protected by biometric locks and armed guards. The viral samples are stored in Sub-Level 5, in a vault that requires dual authorization."

"Dual authorization?"

"Klaus and Wilhelm von Hess both have to be present for access." Eva's voice was hollow with the impossibility of their situation. "Ben, we're talking about breaking into a fortress, penetrating multiple security layers, and somehow forc-ing the two most dangerous men in the Phoenix Order to cooperate."

Ben studied the blueprints, his tactical mind working through the challenges. "What about your access codes? Your biometric data?"

"Probably still active. Klaus is arrogant enough to believe that I'd never return voluntarily." Eva's scientific mind was already working through the implications. "But even with my codes, we'd need to reach the facility, bypass external security, and access the laboratory without triggering alarms."

"One problem at a time." Ben's voice carried the confidence of someone who'd spent years solving impossible problems. "First, we need to get to Bavaria. Then we worry about infiltration."

Eva nodded, but her expression remained troubled. "Ben, there's something else. Something I discovered while decoding the lullaby."

"What?"

"The counter-virus has a critical limitation." Eva's honesty was brutal. "It's designed to neutralize the Lethe Virus, but only the original version. If Klaus has made significant modifications, if the viral structure has evolved beyond my mother's projections, the counter-virus might not be effective."

Ben felt his hope faltering. "How likely is it that the virus has been modified?"

"Very likely. Klaus is paranoid about security, about countermeasures. He's probably updated the viral genome multiple times since my mother's death."

"Then we need access to the current specifications."

"Which means we need Klaus's personal research files." Eva's voice carried the weight of impossibly complex thoughts. "Ben, we're not just talking about stealing viral samples. We're talking about breaking into my father's private laboratory, accessing his encrypted research, and updating the counter-virus formula in real-time."

"How long would that take?"

"Assuming I could access his files, assuming the modifications aren't too extensive, assuming I could synthesize the updated counter-virus without complications—twelve hours minimum."

Ben calculated the timeline. "We have thirty-one hours before release. Twelve hours to update the formula, six hours to synthesize and test the counter-virus, leaves us thirteen hours to deploy it globally."

"Deployment is another problem entirely." Eva's scientific mind was cataloging every obstacle. "The counter-virus would need to be distributed through the same systems the Phoenix Order plans to use. We'd need to access airport ventilation systems, coordinate with authorities, and ensure global coverage within hours."

"Is it possible?"

Eva was quiet for a long moment, her brilliant mind working through the logistics. "Theoretically, yes. The counter-virus is designed to be airborne, stable, and fast-acting. If we could access the Phoenix Order's deployment infrastructure, we could theoretically turn their own weapon against them."

"But?"

"But it would require perfect execution, perfect timing, and perfect cooperation from institutions that may be compromised." Eva's voice carried the weight of their impossible situation. "Ben, we're talking about the most complex covert operation in human history, executed by two people with no resources, no allies, and no margin for error."

Ben moved to the window, staring out at Vienna's ancient streets. Somewhere in the city, Phoenix Order operatives were closing in on them. In Bavaria, Klaus was preparing for the final phase of his genocidal plan. And in thirty-one hours, twelve airports would become ground zero for humanity's extinction.

"Eva," he said quietly, "I need to ask you something."

"What?"

"Are you willing to die for this?"

Eva's answer was immediate and absolute. "Yes."

"Are you willing to kill for it?"

Eva hesitated, her scientific training warring with moral necessity. "If it means saving four billion lives? Yes."

"Then we're going to Bavaria." Ben's voice carried the conviction of someone who'd made an irrevocable choice. "We're going to break into your father's laboratory, steal his research, synthesize the counter-virus, and deploy it before the Phoenix Order can complete their plan."

"And if we fail?"

"Then we die knowing we tried everything." Ben's eyes met hers, and Eva saw something there that made her breath catch. "But we're not going to fail.

We're going to succeed because we have something the Phoenix Order doesn't understand."

"What's that?"

"We have love." Ben's voice was steady with absolute certainty. "Love for humanity, love for the future, love for each other. That's stronger than ideology, stronger than hate, stronger than fear."

Eva moved to him, her hands finding his face. "Ben, I need you to understand something. If we go to Bavaria, if we face Klaus and the Phoenix Order, one of us might not survive."

"I know."

"I need you to promise me something." Eva's voice was steady despite the tears in her eyes. "If something happens to me, if Klaus captures me or I'm killed, you can't try to save me. You have to complete the mission."

"Eva—"

"Promise me." Her voice was fierce with determination. "Promise me that humanity's survival comes first. That four billion lives matter more than one."

Ben was quiet for a long moment, his hands covering hers. "I promise. But I need you to promise me something too."

"What?"

"That you'll remember what we've found together. That no matter what happens, no matter what Klaus says or does, you'll remember that love is stronger than hate. That hope is stronger than despair."

"I promise."

They held each other in the Vienna safe house, two people about to attempt the impossible, knowing that their love might be all that stood between humanity and extinction.

The lullaby had revealed its secret. The counter-virus was possible. But turning possibility into reality would require them to face the most dangerous man in the world—Eva's own father.

"When do we leave?" Eva asked.

"Now." Ben's voice carried the urgency of their situation. "Every hour we delay makes success less likely."

As they gathered their equipment and prepared for the journey that would determine humanity's fate, Eva hummed her mother's lullaby one last time. The melody that had once been a simple childhood comfort had become the key to salvation.

The Phoenix Order had made a crucial mistake. They'd assumed that killing Eva's mother would end the threat to their plans.

Instead, they'd ensured that a mother's love would transcend death itself.

The lullaby would save the world.

Or they would all die trying.

But the choice was made. The path was clear.

In thirty hours, the world would learn whether love or hate would triumph.

The kill switch had been discovered.

Now came the hardest part.

Using it.

The journey to Bavaria began.

And with it, the final battle for humanity's soul.

The Phoenix Order thought they held all the cards.

They were about to learn that a mother's love was the most powerful weapon of all.

The hunt was ending.

The real war was about to begin.

# Chapter 18: The Global Hunt

The first thing Eva noticed was the silence.

The Vienna coffee shop, which had been buzzing with morning conversation just moments before, had gone completely quiet; every patron stared at the wall-mounted television with the kind of horrified fascination usually reserved for natural disasters. Eva followed their gaze and felt her blood turn to ice.

Her own face filled the screen, a professional headshot from her university days, followed immediately by Ben's faculty photo from Berlin. The news anchor's voice cut through the stunned silence with clinical precision.

"International manhunt underway for Dr. Eva Richter and Dr. Benjamin Carter, wanted in connection with a bioterrorism plot targeting major European cities. Authorities describe them as extremely dangerous and advise the public not to approach if spotted."

Eva's hands trembled as she reached for her coffee, trying to maintain the appearance of normalcy while her world collapsed around her. Beside her, Ben's face remained perfectly calm, but she could see the tension in his shoulders, the way his eyes never stopped scanning for threats.

"We need to move," he said quietly, his voice pitched low enough to avoid attention. "Now."

But it was too late. A middle-aged woman at the next table had already pulled out her phone, her finger hovering over what was obviously an emergency number. Her eyes darted between Eva and the television screen, confirmation growing stronger with each glance.

"Excuse me," the woman said, her voice carrying the uncertain authority of someone doing their civic duty. "Aren't you..."

Eva's scientific mind went blank, but Ben's training kicked in immediately. He smiled with the easy charm of someone accustomed to deflecting attention, his voice taking on the slight accent of a German tourist.

"I'm sorry, do we know each other? My wife and I are just visiting from Munich." He gestured to Eva with casual affection. "People always say she looks like someone famous. Very flattering, yes?"

The woman hesitated, her certainty wavering in the face of Ben's confident performance. But Eva could see other patrons beginning to take notice, phones being discreetly activated, the subtle shift in body language that indicated growing suspicion.

"We should go," Ben continued, standing and helping Eva to her feet. "Long day of sightseeing ahead."

They walked toward the exit with deliberate casualness, every instinct screaming at them to run while their rational minds demanded they maintain the facade. Behind them, Eva could hear the woman speaking urgently into her phone, describing their appearance, their direction of travel.

The moment they stepped onto the Vienna street, Eva realized the true scope of their predicament. Every face turned toward them with a hint of recognition. Every security camera became a weapon pointed at their freedom. Every police siren in the distance carried the promise of capture.

"How is this possible?" Eva asked, her voice barely audible above the city noise. "How did they get our photographs? How did they craft such a complete narrative?"

"The Phoenix Order has been planning this for months," Ben replied, guiding her through the crowd with practiced surveillance awareness. "They needed a cover story for the bioweapons attack, scapegoats to blame when four billion people start dying. We're perfect for the role."

Eva's phone buzzed with news alerts, each one more damning than the last. Social media was exploding with their images, amateur investigators dissecting their backgrounds, and conspiracy theories spreading like wildfire. The narrative was elegant in its simplicity: two rogue scientists planning to unleash biological warfare on an unsuspecting world.

"Look at this," Eva said, showing Ben her screen. "They're calling it Operation Prometheus. They're saying we stole classified bioweapons research, that we've been planning this attack for years."

Ben's expression darkened as he read the details. "They've turned your protein synthesis work into evidence of weapons development. My intelligence background makes me the perfect mastermind. The psychological profiles, the Syrian operation, and even our academic positions. Everything we are has been weaponized against us."

A police car turned the corner ahead of them, its occupants scanning the crowd with systematic intensity. Ben pulled Eva into a shop doorway, using the architectural shadows to break their silhouettes.

"We can't travel openly anymore," he said. "Every border crossing, every train station, every airport will have our photographs. We're international fugitives now."

Eva felt the weight of impossibility settling on her shoulders. "How do we get to Bavaria? How do we reach Klaus's laboratory when every law enforcement officer in Europe is hunting us?"

"The underground," Ben replied. "Smuggling networks, criminal transportation, the kind of people who move human cargo without asking questions."

"That could take days. We have twenty-nine hours before the virus release."

"Then we better hope the criminal underground is efficient."

They moved through Vienna's streets like ghosts, using Ben's tradecraft to avoid surveillance cameras and patrol routes. But Eva could feel the net tightening with each passing hour. Their faces were everywhere now, displayed on phones, tablets, and television screens. The technology that connected the modern world had become a web designed to trap them.

The contact Ben had mentioned was located in Vienna's Prater district, operating from a seemingly legitimate shipping company that specialized in "discreet transportation services." The man who met them was exactly what Eva expected from central casting: heavy-set, suspicious, and clearly more interested in money than morality.

"You're expensive cargo," he said in accented English, studying their photographs on his phone. "Very expensive. The kind that brings attention I don't need."

"How much?" Ben asked, cutting through the posturing.

"Fifty thousand euros. Cash. No questions, no guarantees, no refunds if you're caught."

Eva's heart sank. They had maybe three thousand euros between them, the emergency funds Ben had hidden in his go-bag. Fifty thousand might as well have been fifty million.

"We don't have that kind of money," she said.

The smuggler shrugged with practiced indifference. "Then you don't have a passage. Market forces, you understand. Supply and demand."

Ben was quiet for a moment, his analytical mind working through alternatives. "What if we could provide something more valuable than money?"

"Like what?"

"Information. Intelligence about government operations, security protocols, the kind of knowledge that could be worth millions to the right buyers."

The smuggler's eyes sharpened with interest. "What kind of information?"

"CIA operational procedures. European intelligence networks. Counter-terrorism protocols." Ben's voice carried the confidence of someone offering genuine value. "The kind of intelligence that governments kill to protect."

Eva stared at Ben in shock, realizing he was offering to betray every secret he'd sworn to protect. The intelligence that had been beaten into him through years of training, the classified knowledge that represented thousands of lives and billions of dollars in security infrastructure.

"Ben," she whispered, "you can't."

"I can and I will," he replied, his voice steady with absolute conviction. "Four billion lives are worth more than operational security."

The negotiation that followed was swift and brutal. Ben traded secrets that had taken him years to acquire for passage to Bavaria. These burning bridges could never be rebuilt, destroying relationships that had taken decades to forge. Eva watched him sacrifice everything he'd once been for the chance to save the world.

"It's done," the smuggler said finally. "You leave in two hours. Cargo container to Salzburg, then overland to your destination."

As they waited in the shipping company's warehouse, surrounded by crates and containers that held their own dark secrets, Eva felt the weight of what Ben had sacrificed settling between them.

"I'm sorry," she said quietly. "I'm sorry you had to do that."

"I'm not." Ben's voice was steady despite the magnitude of his betrayal. "Those secrets were just information. What we're fighting for is life itself."

Eva's phone buzzed with another news alert, this one more chilling than the rest. Wilhelm von Hess had given a press conference, his aristocratic features radiating calm authority as he described the manhunt in terrifyingly personal terms.

"Dr. Richter and Dr. Carter represent the most dangerous form of terrorism," von Hess was saying, his voice carrying the weight of absolute conviction. "They have perverted science itself, turning humanity's greatest tools into weapons of mass destruction. Make no mistake: these individuals will stop at nothing to achieve their genocidal goals."

Eva watched her father's mentor transform her into a monster with surgical precision. Every word was chosen for maximum impact, every phrase designed to ensure that no one would help them, that no one would question the necessity of their elimination.

"He's making it personal," Ben observed. "Von Hess is taking direct control of the hunt."

"What does that mean?"

"It means we're not just facing the Phoenix Order anymore. We're facing Wilhelm von Hess himself, and he's the most dangerous man in the world."

The cargo container that would carry them to Bavaria was cramped, airless, and filled with legitimate shipping materials that provided minimal cover. As Eva settled into the narrow space between crates of industrial equipment, she realized that they were crossing a line from which there could be no return.

"Ben," she said as the container doors began to close, "what if we're wrong? What if the counter-virus doesn't work? What if we fail and four billion people die anyway?"

"Then we fail knowing we tried everything," Ben replied, his voice carrying absolute conviction. "But we're not going to fail. We're going to succeed because we have something the Phoenix Order doesn't understand."

"What's that?"

"We have each other. We have love. And sometimes, that's enough to move mountains."

The container doors sealed shut, plunging them into darkness as the truck began its journey toward the Bavarian Alps. Outside, the global manhunt continued with mechanical precision. Television screens around the world displayed their photographs, police forces coordinated international searches, and ordinary citizens became unwitting hunters in the Phoenix Order's web.

But inside the container, traveling through the darkness toward an uncertain fate, Eva and Ben had something the Phoenix Order could never manufacture or buy or steal.

They had hope.

The global hunt had made them enemies of every government, fugitives from every law enforcement agency, targets for every bounty hunter and amateur detective who wanted their fifteen minutes of fame.

But it had also stripped away every pretense, every safety net, every comfortable illusion. They were down to the essential elements: love, determination, and the knowledge that four billion lives hung in the balance.

The Phoenix Order had made a critical mistake. They'd assumed that global persecution would break them, that international isolation would force them to surrender.

Instead, they'd created something far more dangerous than two individual fugitives.

They'd created martyrs.

And martyrs, as history had proven time and again, were capable of miracles.

The truck rolled through the Austrian countryside, carrying its human cargo toward the most dangerous place on Earth. In twenty-seven hours, the Phoenix Order would release their weapon of mass extinction.

But they were about to discover that some people were willing to sacrifice everything to prevent that future.

The global hunt was intensifying with each passing hour.

But the hunted were no longer running.

They were coming home.

And they were ready to fight for humanity's soul.

The final battle was approaching.

And love, as always, would be the deciding factor.

The Phoenix Order controlled governments, armies, and the instruments of global surveillance.

But Eva and Ben controlled something more powerful.

They controlled hope.

And sometimes, hope was enough.

The darkness of the container enveloped them as they traveled toward their destiny.

But they were no longer afraid of the dark.

They were ready to become the light.

The hunt continued.

But now, the hunters had become the hunted.

And the world was about to discover which side had chosen correctly.

Twenty-seven hours remained.

The countdown to humanity's salvation had begun.

# Chapter 19: The Midpoint Revelation

The cargo container had become a claustrophobic prison of steel and darkness, filled with the mechanical rhythm of the truck's engine and the scent of industrial lubricants. Eva pressed her back against the cold metal wall, feeling every bump and turn as they traveled deeper into the Austrian countryside. Twenty-four hours remained until the Christmas Eve deadline, and each passing mile brought them closer to what felt increasingly like a suicide mission.

Ben sat across from her in the narrow space between shipping crates, his laptop balanced on his knees, the screen's blue glow providing the only light in their metal tomb. For the past three hours, he'd been working through encrypted channels, using the criminal smuggler's satellite connection to penetrate intelligence networks that should have been impossible to access.

"Talk to me," Eva said, her voice barely audible above the engine noise. "What are you finding?"

Ben's expression was grim in the laptop's pale light. "It's worse than we thought. Much worse."

Eva's stomach clenched with dread. "How much worse?"

"The Phoenix Order hasn't just planned a coordinated attack. They've already deployed." Ben turned the laptop toward her, showing a map of the world dotted with red markers that made Eva's blood run cold. "Forty-seven carriers in position across six continents. Not twelve airport attacks, Eva. Forty-seven simultaneous releases."

Eva stared at the map, her scientific mind struggling to process the implications. Red dots marked major cities across the globe: New York, London, Paris, Tokyo, Sydney, São Paulo, Mumbai, Lagos, Cairo, Moscow, Beijing, Mexico City. Every major population center, every critical transportation hub, every strategic location that could ensure maximum viral distribution.

"My God," she whispered. "The scope is massive."

"It gets worse." Ben pulled up another file, his voice hollow with the weight of discovery. "These aren't suicide bombers or fanatic volunteers. Most of the carriers don't even know they're infected."

"What do you mean?"

"The Phoenix Order has been conducting 'medical trials' for months. Fake vaccine programs, experimental treatments, and routine blood work at compromised medical facilities. They've infected their own operatives with a dormant version of the Lethe Virus."

Eva's hands flew to her mouth in horror. "They're walking biological weapons."

"Exactly. The carriers are living normal lives, traveling to their assigned locations, completely unaware that they're about to trigger a global apocalypse." Ben's analytical mind was processing the elegant cruelty of the plan. "Christmas Eve, 1200 GMT. Simultaneous activation through an encoded radio signal that triggers the dormant virus."

"Activation?"

"The Lethe Virus has been engineered with a remote trigger. A specific electromagnetic frequency that causes the dormant pathogen to become active and contagious." Ben's voice carried the professional detachment of someone discussing a technical marvel that happened to be designed for mass murder. "The carriers will begin showing symptoms within hours, but by then they'll have infected hundreds of people in airports, shopping centers, hotels, and tourist attractions."

Eva studied the deployment map, her scientific training allowing her to visualize the infection patterns. "Global saturation within seventy-two hours. Billions exposed before anyone realizes what's happening."

"And that's just the first wave." Ben opened another file, revealing infection modeling that made Eva's chest tighten with despair. "The Phoenix Order has spent years identifying super-spreader locations, optimal timing for maximum exposure, even weather patterns that will enhance airborne transmission."

The scope of the operation was breathtaking in its thoroughness. Every variable had been accounted for, every contingency planned, every possible countermeasure anticipated. The Phoenix Order hadn't just created a weapon of mass destruction. They'd engineered the perfect plague.

"Ben," Eva said quietly, "we can't stop this. Forty-seven simultaneous attacks across six continents? Even if we could identify every carrier, even if we could convince authorities to take action, the logistics are impossible."

"I know." Ben's honesty was brutal. "Which means we need to change our strategy."

"To what?"

"We stop trying to prevent the attack and focus on surviving it." Ben's voice carried the cold logic of someone who'd spent years analyzing impossible scenarios. "We can't stop the virus from being released, but we might be able to limit its impact."

Eva felt that something fundamental had shifted in her understanding of their mission. For days, she'd been focused on preventing the Phoenix Order's attack, on stopping the virus before it could be deployed. But Ben was right. That strategy was no longer viable.

"You're talking about the counter-virus," she said.

"I'm talking about turning the Phoenix Order's own distribution network against them." Ben's eyes met hers, and Eva saw a desperate plan forming in his analytical mind. "If we can synthesize your mother's counter-virus, if we can deploy it through the same systems they're using for the attack, we might be able to neutralize the Lethe Virus before it reaches lethal concentrations."

"That's insane. The timing would have to be perfect, the distribution simultaneous, the counter-virus effective against whatever modifications Klaus has made." Eva's scientific training catalogued every obstacle. "We'd need laboratory facilities, synthesis equipment, samples of the current virus, and access to the Phoenix Order's deployment infrastructure."

"All of which are located in one place," Ben said quietly.

"Klaus's laboratory." Eva's voice was hollow with realization. "You're suggesting we break into the most secure facility in Europe, steal viral samples, synthesize a counter-virus we've never tested, and somehow hijack a global distribution network."

"I'm suggesting it's our only chance to save four billion lives."

The truck hit a pothole, causing them both to bounce against the container walls. Eva used the momentary distraction to study Ben's face in the laptop's glow, seeing the same desperate determination that had driven her to steal her father's files, to risk everything for a chance to prevent genocide.

"Even if we could access the laboratory, even if we could synthesize the counter-virus, how would we deploy it globally?" Eva's scientific mind demanded practical answers. "The Phoenix Order's distribution network is designed for their operatives, not for external access."

"Klaus will have command and control systems," Ben replied. "Master protocols that can coordinate the entire operation. If we can access those systems, we can potentially reverse the signal, turn the activation frequency into a neutralization frequency."

"You're assuming the counter-virus can be triggered the same way as the original pathogen."

"Your mother was brilliant. If she designed the counter-virus to neutralize the Lethe Virus, she would have made it compatible with their delivery systems." Ben's voice carried more hope than certainty. "She would have anticipated this exact scenario."

Eva closed her eyes, trying to recall every detail of her mother's lullaby, every nuance of the genetic code hidden in childhood memories. The melody was complex and sophisticated, but was it sophisticated enough to account for remote activation protocols?

"Maybe," she said finally. "But we'd still need to reach the laboratory, access Klaus's personal research, update the counter-virus formula for any modifications he's made, synthesize it in sufficient quantities, and deploy it within hours of the original attack."

"How long would synthesis take?"

"Assuming perfect conditions, no complications, and industrial-grade equipment? Twelve hours minimum." Eva's honesty was brutal. "That's assuming I could decode Klaus's modifications immediately, assuming the synthesis process works perfectly, assuming we don't encounter any unexpected chemical interactions."

Ben calculated the timeline in his head. "Twenty-four hours until activation. Twelve hours for synthesis. That gives us twelve hours to infiltrate the facility, access the research, and prepare for deployment."

"And escape?"

Ben was quiet for a long moment. "I don't think escape is part of this plan."

The words hung between them in the container's darkness, carrying the weight of absolute finality. Eva had known this moment would come, had understood from the beginning that stopping the Phoenix Order might require the ultimate sacrifice. But hearing it spoken aloud made the reality inescapable.

"A suicide mission," she said quietly.

"A mission that might save humanity." Ben's voice was steady despite the magnitude of what he was proposing. "Eva, I need you to understand something. The laboratory is a fortress. Klaus will be expecting us. The security is military-grade, the location is isolated, and the Phoenix Order will have every advantage."

"But?"

"But it's also the only place where we can access everything we need. The viral samples, the synthesis equipment, the command and control systems, Klaus's personal research. Everything required to deploy the counter-virus is located in that facility."

Eva opened her eyes, studying Ben's face in the laptop's glow. "You've been planning this since Vienna, haven't you? This was always the endgame."

"I've been hoping we'd find another way. But the intelligence I'm seeing, the scope of their operation, the impossibility of stopping forty-seven simultaneous attacks..." Ben's voice trailed off. "Sometimes the only choice is between heroic failure and cowardly success."

"Meaning?"

"We could run. Disappear into the European underground, find some remote corner of the world, and wait for the Phoenix Order to complete their genocide. We'd survive, but four billion people would die."

"Or?"

"We infiltrate Klaus's laboratory, knowing we probably won't survive, but with a chance to deploy the counter-virus before the Lethe Virus reaches lethal concentrations."

Eva was quiet for a long moment, her mind working through the implications. Everything she'd worked for, everything she'd believed in, everything she'd dreamed of for the future, reduced to a single impossible choice.

"There's something else," Ben said, his voice carrying a note of reluctant honesty. "Something I haven't told you about the laboratory's location."

"What?"

"It's not just a research facility. It's Klaus's home. The monastery where you grew up, where your mother lived, where your childhood memories were formed." Ben's eyes met hers. "Eva, we're asking you to return to the place where your family was destroyed, to face the man who killed your mother, to risk everything in the building where you learned to trust and love."

Eva felt tears stinging her eyes as the emotional weight of their mission crystallized. "He's counting on that, isn't he? Klaus knows that returning home will be the hardest thing I've ever done."

"Probably. He knows your psychological profile, your emotional triggers, and your vulnerabilities. The monastery isn't just his fortress. It's his psychological weapon against you."

"And you think I can handle it?"

"I think you're the strongest person I've ever met." Ben's voice carried absolute conviction. "I think you've already survived the worst thing Klaus could do to you. He destroyed your family, perverted your research, and turned your father into a monster. What's left for him to take?"

Eva considered the question, feeling something shifting deep in her chest. Klaus had already taken her mother, her childhood innocence, and her faith in her father's goodness away from her. The laboratory held memories of loss and betrayal, but it also had something else.

"My mother's laboratory," she said suddenly. "Her private research space in the monastery's east wing. Klaus never touched it after she died. If there are additional notes, backup research, anything that might help with the counter-virus synthesis..."

"It could give us the edge we need," Ben finished. "Your mother's final gift to humanity."

Eva nodded, feeling a strange sense of completion settling over her. The monastery had been the site of her greatest loss, but it might also be the site of her greatest triumph. Her mother's love, encoded in a lullaby and hidden in childhood memories, might be the key to humanity's salvation.

"Ben," she said quietly, "I need to ask you something."

"What?"

"Why are you doing this? Why are you willing to die for people you've never met, for a world that's branded you a terrorist?"

Ben was quiet for a long moment, his fingers absently tracing patterns on the laptop's keyboard. "Do you remember what I taught in my Damascus lecture? About the banality of evil?"

"Of course."

"I spent three years thinking about that concept, about how ordinary people become complicit in extraordinary evil through inaction, through moral cowardice, through the failure to take a stand when it matters most." Ben's voice carried the weight of three years of self-reflection. "The Phoenix Order is counting on that. They're counting on people being too frightened, too overwhelmed, too concerned with their own survival to resist."

"And you're not?"

"I'm terrified. But I'm more terrified of living in a world where love loses to hate, where courage yields to cowardice, where the future belongs to people like Klaus and von Hess." Ben's eyes met hers. "Eva, you've shown me what it means to fight for something bigger than yourself. You've reminded me that some things are worth dying for."

"Like what?"

"Like the chance for children to grow up in a world where scientists work to heal rather than kill. Like the possibility that human beings can choose compassion over cruelty. Like the hope that love is stronger than ideology."

Eva felt her chest tighten with emotion. "Ben, if we do this, if we go to the laboratory, there's a good chance we won't survive."

"I know."

"But if we don't do this, if we let the Phoenix Order succeed, four billion people will die."

"I know that too."

Eva moved closer to him in the cramped space, her hands finding his face in the darkness. "Then we're really going to do this. We're really going to break into Klaus's laboratory, steal his research, synthesize the counter-virus, and try to save the world."

"Together," Ben said, his voice steady with absolute certainty.

"Together," Eva agreed.

The truck hit another pothole, reminding them of their precarious situation. They were international fugitives traveling in a smuggler's container, heading toward the most dangerous place on Earth, carrying the weight of humanity's future on their shoulders.

But they were also something else. They were two people who'd found each other in the darkness, who'd chosen love over fear, who'd decided that some things were worth any sacrifice.

"Eva," Ben said quietly, "there's something I need to tell you."

"What?"

"I love you. Not just because you're brilliant, not just because you're brave, not just because you're fighting to save the world. I love you because when I look at you, I see everything I want to be. Everything I want the world to become."

Eva's tears were flowing freely now, hot against her cheeks in the container's cold air. "I love you, too. I love your courage, your conscience, your willingness to sacrifice everything for people you'll never meet. I love that you've shown me what it means to choose hope over despair."

They held each other in the darkness, two people about to attempt the impossible, knowing that their love might be all that stood between humanity and extinction.

The truck continued its journey through the Austrian countryside, carrying them toward the Bavarian Alps and the final confrontation with the Phoenix Order. Outside, the world continued its oblivious existence, unaware that in twenty-three hours, everything would change.

But inside the container, traveling through the darkness toward an uncertain fate, Eva and Ben had found something that the Phoenix Order could never understand or destroy.

They had found each other. They had found love. And they had found the courage to fight for humanity's soul.

The midpoint of their journey had revealed the true scope of their enemy's power and the impossibility of their mission. The Phoenix Order controlled governments, armies, and the instruments of global destruction.

But Eva and Ben controlled something more powerful.

They controlled hope.

And sometimes, hope was enough to move mountains.

The countdown to humanity's salvation continued.

Twenty-three hours remained.

The final battle was approaching.

And love, as always, would be the deciding factor.

The truck rolled through the night, carrying its human cargo toward destiny.

The Phoenix Order thought they had won.

They were about to discover that some battles were worth fighting regardless of the odds.

Some causes were worth dying for.

And some loves were strong enough to save the world.

The hunt was ending.

The real war was about to begin.

And in a monastery hidden in the Bavarian Alps, a father and daughter would face each other one final time.

The midpoint revelation was complete.

The end game had begun.

# Chapter 20: The Approach

The Alpine Forest was a cathedral of darkness, its whispers carrying warnings.

Eva crouched behind a massive pine tree, her breath forming small clouds in the December cold as she watched the helicopter's searchlight sweep across the mountainside below them. The mechanical insect had been hunting for three hours now, its thermal imaging sensors probing every shadow, every cave, every potential hiding place where two fugitives might seek shelter.

"Third pass in the last hour," Ben whispered, his voice barely audible above the wind. "They're not just searching randomly. They know we're here."

Eva nodded, her scientific mind already processing the implications. The Phoenix Order's pursuit had evolved beyond simple manhunting into something approaching omniscience. Every route they'd taken, every decision they'd made, every survival choice that should have been unpredictable had somehow been anticipated.

The cargo container had delivered them to a truck stop outside Salzburg, where their criminal smuggler had unceremoniously dumped them at 0347 hours with a curt warning about police roadblocks and increased surveillance. From there, they'd stolen a motorcycle. They attempted to reach the Bavarian border through mountain roads that should have been invisible to satellite tracking.

Instead, they'd encountered roadblocks that seemed positioned specifically for their route, patrol helicopters that appeared at precisely the wrong moments, and surveillance drones that tracked their movement with algorithmic precision. The Phoenix Order wasn't just hunting them anymore. They were herding them, using artificial intelligence and predictive modeling to anticipate their choices before they made them.

"How are they doing this?" Eva asked, her breath visible in the cold air. "How are they staying ahead of us when we're making decisions in real time?"

Ben's expression was grim as he studied the helicopter's search pattern. "Behavioral modeling. They're feeding our psychological profiles into AI systems that can predict human decision-making with terrifying accuracy."

"But we're being unpredictable. We're avoiding main roads, changing direction randomly, making choices that contradict our normal patterns."

"Are we?" Ben's voice carried the weight of uncomfortable realization. "Think about it, Eva. Every choice we've made has been logical given our circumstances. Take the mountain route to avoid surveillance. Steal transportation when public travel becomes impossible. Seek high ground for tactical advantage. We think we're being unpredictable, but we're actually following classic evasion protocols."

Eva felt a chill that had nothing to do with the cold of the Alps. "They're not predicting our specific choices. They're predicting the logical framework that drives our choices."

"Exactly. Klaus knows how you think, how you approach problems, and how you make decisions under pressure. My CIA training gives me predictable tactical instincts. Combined with AI modeling that can process thousands of variables simultaneously..." Ben's voice trailed off as the implications sank in.

The helicopter completed another sweep and banked toward the east, its searchlight probing a different section of the mountainside. But Eva knew it would be back. The Phoenix Order had unlimited resources, unlimited patience, and algorithms that were learning their behavioral patterns with each passing hour.

"We need to do something genuinely unpredictable," she said. "Something that contradicts our training, our instincts, our logical decision-making process."

"Like what?"

Eva studied the terrain around them, her mind working through options that violated every survival principle Ben had taught her. The Alpine Forest stretched in all directions, offering both concealment and exposure, depending on how it was used.

"The monastery is twelve kilometers northeast of here," she said, pointing toward a mountain peak barely visible in the darkness. "The logical approach would be to maintain concealment, move during darkness, avoid all populated areas, and approach from the most defensible direction."

"Which would be?"

"From the south, using the forest cover, approaching through the old smuggling routes that connect to the wine cellars." Eva's voice carried the certainty of childhood knowledge. "Klaus would expect that approach. He'd position his security to cover those traditional infiltration routes."

Ben studied her face in the dim starlight. "So, what's the unpredictable alternative?"

"We go straight through the village. In broad daylight. Using the main road."

"Eva, that's suicide. We're international fugitives. Our faces are on every television screen, every police bulletin, every smartphone app in Central Europe."

"Exactly. Which is why it's the last thing they'd expect." Eva's scientific mind was processing the variables with cold, logical precision. "Klaus's behavioral models assume we'll act like fugitives, like trained operatives, like people trying to avoid capture. But what if we don't?"

Ben was quiet for a moment, his tactical training warring with Eva's unconventional logic. "Walk me through it."

"The village of Sankt Georgen sits directly between our current position and the monastery. The population may be three hundred people, mostly elderly, heavily Catholic, the kind of community where strangers are noticed but not necessarily reported if they seem harmless."

"And we don't seem harmless. We seem like international terrorists."

"Do we?" Eva's voice carried a note of challenge. "Or do we seem like two exhausted travelers who've been hiking in the mountains? Two people who need food, shelter, and basic human kindness?"

Ben studied her face, seeing the plan forming behind her brilliant eyes. "You're talking about hiding in plain sight."

"I'm talking about becoming different people. Not fugitives trying to avoid capture, but travelers in need of help." Eva's voice gained strength as the concept crystallized. "Klaus's algorithms are designed to track evasive behavior, paranoid decision-making, the movement patterns of people trying to hide. But they're not designed to track ordinary human vulnerability."

The more Eva explained her reasoning, the more Ben realized she might be right. The Phoenix Order's surveillance network was optimized for hunting dangerous criminals, not for identifying exhausted hikers seeking assistance from rural villagers.

"It's insane," he said finally. "Completely against every principle of operational security."

"Which is exactly why it might work." Eva's smile was barely visible in the darkness. "Sometimes the most dangerous thing you can do is the thing that seems safest."

They spent the next hour modifying their appearance with materials scavenged from the forest. Mud and pine sap to dirty their clothes and faces, giving them the appearance of people who'd been camping rough for days. Ben used his knife to create tears in their clothing that suggested hard travel rather than deliberate disguise. Eva braided pine needles into her hair, creating the kind of natural camouflage that hikers might adopt for practical rather than tactical reasons.

"Remember," Ben said as they prepared to leave their hiding place, "we're not international fugitives. We're Klaus and Greta Mueller from Innsbruck, attempting a winter hiking circuit that proved more challenging than expected."

"Lost our equipment in a river crossing," Eva continued, falling naturally into the cover story they'd constructed. "Need food, warm clothing, and directions to the nearest transportation."

"And if someone recognizes us from the news reports?"

"We laugh it off. Point out how much the stress of mountain hiking can change someone's appearance. Make jokes about how everyone looks like a criminal when they haven't showered in days."

Ben nodded, though every instinct screamed against this plan. They were placing their lives in the hands of strangers, trusting that human compassion would prevail over civic duty, and gambling that the Phoenix Order's technological sophistication would be overcome by simple human kindness.

The descent from their hiding place took two hours, following game trails and seasonal streams that brought them to the outskirts of Sankt Georgen just as the December sun was beginning to rise over the Alps. The village was picture-perfect in its Alpine charm, featuring wooden chalets with flower boxes and a white church with a distinctive onion dome. These narrow streets had probably remained unchanged for centuries.

"It's beautiful," Eva whispered, genuine emotion coloring her voice. "I remember coming here with my mother when I was small. She used to buy bread from the bakery and take me to see the Christmas markets."

Ben caught the note of nostalgia and pain in her voice, reminding him that this wasn't just a tactical operation for Eva. This was a homecoming to a childhood that had been destroyed by her father's choices.

"Stay focused," he said gently. "We can't afford to be sentimental."

"I know. But it helps to remember that there was goodness here once. That not everything from my past was corrupted by Klaus's ideology."

They approached the village from the east, following the main road with the deliberate casualness of people who had nothing to hide. Eva's transformation was remarkable. Gone was the tense awareness of a hunted fugitive, replaced by the slightly sheepish demeanor of someone who'd overestimated their hiking abilities.

The first test occurred at the village's edge, where an elderly man was tending to his chickens in his yard. He looked up as they approached, his weathered face showing the kind of curiosity that rural people displayed toward strangers.

"Guten Morgen," Eva called out, her voice carrying just the right note of embarrassed exhaustion. "We're wondering if you might direct us to a telephone. We've had some difficulties with our hiking expedition."

The man studied them for a moment, taking in their muddy clothing and disheveled appearance. "Americans?"

"Austrian," Ben replied, his accent carefully modulated to suggest Innsbruck rather than Berlin. "Though my wife has been practicing her English with tourists, so perhaps our German sounds strange."

The elderly man's expression softened, as if with genuine concern. "You look like you've had a rough time of it. The mountains can be unforgiving in winter."

"More unforgiving than we anticipated," Eva admitted, allowing a note of rueful humor to color her voice. "We lost most of our equipment crossing a stream that proved deeper than expected."

What followed was exactly what Eva had predicted. The elderly man, whose name was Johann, insisted on bringing them to his wife, who insisted on providing hot coffee and fresh bread. Within an hour, they were sitting in a warm kitchen, listening to local weather reports, and receiving advice about the best routes to return to civilization.

"The road to Berchtesgaden is clear," Johann's wife, Margarete, explained while refilling their coffee cups. "But the mountain passes are treacherous this time of year. Many accidents, much ice."

"We've learned our lesson about winter hiking," Ben said with appropriate humility. "Is there perhaps a local driver who might give us transportation to the nearest train station?"

"My nephew could help," Johann suggested. "He has a truck and makes deliveries to Salzburg twice a week. For travelers in distress, he would make an exception."

Eva felt a pang of guilt at the deception. These people were showing them genuine kindness, opening their homes to strangers based solely on human compassion. The Phoenix Order's cynical worldview could never account for this kind of spontaneous generosity.

"You're very kind," she said, meaning every word. "We're grateful for your help."

It was while they waited for Johann's nephew that the first sign of danger appeared. Eva was helping Margarete with dishes when she noticed the older woman glancing repeatedly at the small television in the corner of the kitchen.

"Is everything all right?" Eva asked, following Margarete's gaze.

"It's probably nothing," Margarete said, but her voice carried a note of uncertainty. "It's just that there's been news about dangerous people in the area. Terrorists, they say. A man and woman wanted by international police."

# THE PHOENIX STRAIN

Eva's blood went cold, but she forced herself to maintain the appearance of casual curiosity. "How frightening. What kind of terrorists?"

"Scientists, apparently. Bioweapons." Margarete's voice dropped to a whisper, as if discussing such evil might somehow draw it in. "They say the woman looks quite ordinary. Pretty, blonde, could be anyone's daughter."

"Well, I certainly hope the police catch them soon," Eva replied, her voice steady despite her racing heart. "People like that don't belong in civilized society."

Margarete nodded, but Eva could see the woman studying her face with growing attention. The resemblance to the television photographs was probably close enough to raise questions, especially now that the old woman was actively looking for similarities.

"Margarete," Eva said, leaning closer in a gesture of conspiratorial confidence, "may I ask you something?"

"Of course, dear."

"Do you really think I look like a bioterrorist?" Eva's question was delivered with just the right amount of amused incredulity. "Because I have to tell you, after three days of hiking disasters, I feel dangerous enough to qualify."

Margarete's laugh was genuine and relieved. "Oh, my dear girl, forgive an old woman's foolishness. Of course you don't look like a criminal. It's just that the news makes everyone paranoid these days."

The moment passed, but Eva realized how close they'd come to exposure. The Phoenix Order's media campaign was working exactly as intended, turning every civilian into a potential informant, every act of kindness into an opportunity for betrayal.

Johann's nephew arrived an hour later, a cheerful man in his thirties who owned a delivery truck and seemed delighted to help stranded hikers. The ride to Salzburg would take two hours, he explained, but he could drop them at the train station where they could catch connections to anywhere in Austria.

As they loaded into the truck, Eva felt the weight of irony settling on her shoulders. They were traveling toward their destination in comfort and safety, protected by the very people the Phoenix Order claimed to be defending. The ordinary human decency that Klaus dismissed as weakness had become their salvation.

"You know what strikes me about this?" Ben said quietly as the truck wound through mountain roads toward Salzburg.

"What?"

"Klaus's entire ideology is based on the premise that most people are genetically inferior, morally weak, incapable of true courage or compassion. But look at what we've experienced. Strangers opening their homes to people in need, risking their own safety to help travelers in distress."

Eva nodded, feeling the same contradiction. "The Phoenix Order sees humanity as a disease to be cured. But what we're seeing is humanity as it really is, underneath all the fear and propaganda."

"Makes you wonder what Klaus would think if he could see his own neighbors showing kindness to the daughter he's trying to kill."

"I think it would confirm everything he believes about the need to reshape human nature." Eva's voice carried a note of sadness. "Klaus doesn't see kindness as a virtue. He sees it as a weakness that makes people vulnerable to exploitation."

The truck continued its journey through the Alps, carrying them closer to their final destination. Outside the windows, the Bavarian countryside rolled past in picture-perfect perfection, with white snow covering ancient forests, church spires rising from valley floors. This landscape had inspired fairy tales and legends.

But Eva knew that somewhere ahead, hidden in these same mountains, her father was preparing to destroy the world that had created such beauty. The same human impulses that had driven strangers to help them were about to be eliminated as genetic impurities.

"Ben," she said quietly, "whatever happens at the monastery, whatever Klaus says or does, I want you to remember this morning. Remember Johann and Margarete, remember their nephew's willingness to help strangers. Remember that this is what we're fighting for."

"The capacity for kindness?"

"The choice to be kind. Klaus believes that human nature is fixed, that genetics determines behavior, and that some individuals are inherently superior to others. But what we've seen today is proof that people choose who they want to be."

Ben studied her face, seeing the resolution forming behind her intelligent eyes. "You're not just fighting to stop the virus anymore, are you?"

"I'm fighting to prove that my father is wrong about human nature. That love is stronger than hate, that kindness is more powerful than cruelty, that the future belongs to people who choose hope over fear."

The truck crested a ridge, and suddenly the valley containing the monastery was visible below them. Eva's breath caught as she saw the familiar stone walls, the Gothic towers, the terraced gardens where she'd played as a child. Home. The place where her family had lived, loved, and been destroyed.

"There," she said, pointing to the massive structure that seemed to grow from the mountainside itself. "That's where it ends."

Ben followed her gaze, his tactical mind automatically cataloging defensive positions, access routes, and potential vulnerabilities. But he also saw something else. He saw a young woman preparing to return to the site of her greatest trauma, carrying the weight of humanity's future on her shoulders.

"Eva," he said quietly, "are you ready for this?"

"No," she admitted. "I don't think anyone could be ready to face what we're about to face. But I'm willing. And sometimes, that has to be enough."

The truck began its descent into the valley, carrying them toward their final confrontation with the Phoenix Order. Behind them, the village of Sankt Georgen continued its peaceful existence, unaware that it had just provided shelter to the world's most wanted fugitives.

Ahead of them, the monastery lay, waiting with its secrets and dangers, its memories, and horrors. Klaus would be preparing for their arrival, setting traps both physical and psychological, using every advantage of terrain and psychology to ensure their defeat.

But Eva and Ben had something Klaus couldn't account for in his calculations. They had love. They had hope. And they had witnessed the fundamental goodness of humanity in its purest form.

The Phoenix Order believed that human nature needed to be remade through genetic engineering. Eva and Ben had learned that human nature was already perfect, if people were simply given the choice to express their better angels.

The approach to the monastery continued, carrying them toward a confrontation that would determine not only the fate of four billion lives but also the future of human civilization itself.

In eighteen hours, the Phoenix Order would release their weapon of mass extinction.

But in the next few hours, two people who believed in the power of love would attempt something that every rational analysis suggested was impossible.

They would try to save the world.

And they would do it by trusting in the same human kindness that had carried them safely through the Alps.

The final battle was approaching.

But the real victory had already been won in a village kitchen, over coffee and bread, through the simple act of strangers helping strangers.

That was what humanity looked like at its best.

That was what they were fighting to preserve.

And that was why they would not, could not, must not fail.

The monastery grew larger as they approached, its stone walls holding secrets that would determine the future of the human race.

Eva closed her eyes and hummed her mother's lullaby. This genetic code might hold the key to saving the world, hidden in childhood memories of love and loss.

The end game had begun.

The approach was complete.

Time to discover if love really was stronger than hate.

Time to find out if hope could triumph over despair.

Time to learn whether two people with nothing left to lose could move mountains.

The Phoenix Order was waiting.

But so was humanity's future.

And that future was worth any sacrifice, any risk, any price.

The countdown continued.

Seventeen hours remained.

The final battle was about to begin.

# Chapter 21: The Mountain Fortress

T he monastery had been transformed into something that would have made medieval siege architects weep with envy.

Eva lay prone on the snowy ridge overlooking her childhood home, studying the fortress through Ben's military-grade binoculars as afternoon shadows lengthened across the Alpine valley. What she saw below made her stomach clench with a mixture of recognition and horror. The ancient stone walls she'd climbed as a child now bristled with modern surveillance equipment. The gardens where she'd played hide-and-seek had been converted into killing fields with overlapping fields of fire.

"Tell me what you're seeing," Ben whispered beside her, his breath forming small clouds in the December cold.

"Motion sensors along the entire perimeter," Eva replied, her voice hollow with disbelief. "Thermal imaging cameras at regular intervals. What looks like automated weapon systems disguised as decorative stonework." She lowered the binoculars, her hands shaking slightly. "Ben, this isn't just security. This is a military installation."

Ben took the binoculars and conducted his own assessment, his CIA training allowing him to identify threats that Eva's scientific background might have missed. What he saw confirmed her worst fears and added several new ones.

"Guard towers at the four corners, stuffed with sniper teams. Vehicle barriers that can be deployed to block the access road within seconds. Anti-aircraft systems hidden in the bell towers." Ben's voice carried the professional detachment of someone cataloguing an impossibly defended target. "Eva, I count at least forty armed personnel, all carrying automatic weapons, all positioned to create interlocking fields of fire."

"How many people did the monastery used to house?" Ben asked.

"Maybe twenty researchers and support staff. The facility was designed for scientific work, not military operations." Eva's voice cracked slightly as she processed the magnitude of the transformation. "Klaus has turned my childhood home into a fortress."

The scale of the defensive preparations was breathtaking and terrifying. Every approach route had been anticipated and covered. Every potential hiding place had been eliminated or monitored. Every weakness in the ancient architecture had been reinforced with modern technology.

"There's something else," Ben said, adjusting the binoculars' focus. "The modifications aren't random. They're specifically designed to counter infiltration by people with intimate knowledge of the facility."

"What do you mean?"

"The hidden passage you mentioned, the one behind the wine cellar. There's a new guard post positioned to cover that exact entrance. The drainage tunnels have been fitted with motion sensors. Even the old maintenance shafts have been blocked or monitored."

Eva felt a chill that had nothing to do with the Alpine wind. "He's not just defending against intruders. He's defending against me specifically."

"Your father knows how you think, how you move, how you would approach the building. Every childhood hiding place, every secret route, every advantage your knowledge might provide has been neutralized."

The implications were staggering. Klaus hadn't just fortified the monastery against random attack. He'd studied his daughter's behavior patterns, her childhood memories, her psychological tendencies, and designed countermeasures for each potential approach she might attempt.

"It's a trap," Eva said quietly. "Klaus wants us to come here. He's been herding us toward this location, eliminating our alternatives, forcing us into a confrontation on his terms."

Ben studied the defensive positions through the binoculars, his analytical mind working through the tactical implications. "Not just a trap. A psychological weapon. He's using your emotional connection to this place against you."

"How?"

"Think about it. You grew up here. This building contains your happiest childhood memories and your worst traumas. Your mother's laboratory is inside those walls, along with your father's research and the synthesis equipment we need." Ben's voice carried grim understanding. "Klaus has made it impossible for you to complete your mission without returning to the place where your family was destroyed."

Eva closed her eyes, feeling the weight of her father's psychological manipulation. Even from a distance, even through binoculars and defensive fortifications, Klaus was reaching into her mind, using her own memories as weapons against her.

"He knows I'll have to face the laboratory where my mother worked," she said. "The rooms where she was murdered. The places where I trusted him completely before learning what he really was."

"Classic psychological warfare," Ben confirmed. "Force your enemy to fight on terrain that favors your psychological advantages rather than their tactical strengths."

A helicopter appeared over the eastern ridge, its rotors beating the air with mechanical precision. Eva and Ben pressed themselves deeper into the snow as the aircraft began a systematic patrol of the surrounding mountains.

"Third helicopter this hour," Ben observed. "They're not just defending the monastery. They're creating a security perimeter that extends for kilometers in every direction."

Eva watched the helicopter's search pattern, her scientific mind noting the efficiency of their surveillance grid. "Motion sensors, thermal imaging, acoustic detection, seismic monitoring. Klaus has created a detection network that would alert him to any approach long before we reached the building."

"Unless we find a way to mask our signatures."

"How? We don't have military-grade stealth equipment. We don't have electromagnetic jammers or thermal dampening gear. We're two people with basic climbing equipment and whatever supplies we could carry."

Ben was quiet for a long moment, studying the monastery through the binoculars. "Maybe that's our advantage."

"What do you mean?"

"Klaus has prepared for a military-style assault. Professional infiltration techniques, sophisticated equipment, coordinated team tactics. But what if we don't approach like professionals?"

Eva studied his profile, seeing the plan forming behind his intelligent eyes. "You're thinking about the village again. About doing something unpredictable."

"I'm thinking about your childhood. About how a seven-year-old girl might have approached this building when she wanted to explore areas that were off-limits."

Eva felt a spark of recognition. "The old aqueduct system. Roman engineering that predates the monastery by a thousand years."

"Show me."

Eva took the binoculars and traced a route along the mountainside that was invisible unless you knew exactly where to look. "There. About two kilometers east of the main building. The Romans constructed an aqueduct to transport water from mountain springs to the valley floor. Most of it was destroyed over the centuries, but sections remain."

"Including sections that pass under the monastery?"

"Including a section that connects to the building's water supply. My mother showed me the entrance when I was small, back when Klaus trusted me to explore the facility unsupervised." Eva's voice carried the bittersweet weight of childhood memories. "It's not on any architectural plans. It's not marked on any security survey. Most people don't even know it exists."

Ben studied the terrain she'd indicated, his tactical mind evaluating the possibilities. "Could it support adult weight? Could we traverse it without triggering acoustic sensors?"

"Maybe. The stone construction is solid, the passage is narrow but navigable, and it connects directly to the monastery's basement levels." Eva's scientific training demanded honesty. "But it's also dangerous, unstable in places, and if we're discovered inside the tunnel system, we'd be trapped with no escape route."

"How long would the traverse take?"

"Three hours minimum, assuming no complications. The passage is maybe eight hundred meters long, but it's not a straight line. There are branches, dead ends, sections where the ceiling has partially collapsed."

Ben lowered the binoculars, his expression grim with calculation. "Three hours to reach the basement, plus however long it takes to access your mother's laboratory, decode Klaus's modifications to the virus, synthesize the counter-virus, and deploy it through their own distribution network."

"You forgot the part where we somehow avoid being killed by forty armed guards and whatever psychological traps Klaus has prepared."

"One problem at a time." Ben's voice carried the confidence of someone who had spent years solving seemingly impossible problems. "The aqueduct gives us a way inside. Your knowledge of the facility gives us navigation. Your mother's research gives us the counter-virus formula."

"And Klaus's psychological warfare?"

Ben was quiet for a moment, studying Eva's face in the afternoon light. "That's the part we face together. Whatever Klaus says, whatever he does, whatever memories he tries to use against you, we face it as a team."

Eva nodded, but she could feel the fear building in her chest. Returning to the monastery meant confronting every trauma Klaus had inflicted, every trust he'd betrayed, every childhood memory he'd corrupted with his ideology.

"Ben," she said quietly, "what if I can't do it? What if facing Klaus breaks me psychologically? What if I freeze when we need action?"

"Then I'll be there to catch you," Ben replied without hesitation. "Eva, you're the strongest person I've ever met. You've already survived the worst thing Klaus could do to you. What's left for him to take?"

"My sanity. My ability to function under pressure. My capacity to synthesize the counter-virus when billions of lives depend on it."

"Your love for your mother. Your determination to honor her sacrifice. Your commitment to preventing genocide." Ben's voice carried absolute conviction. "Klaus can attack your mind, but he can't change who you fundamentally are. And who you are is someone who chooses love over hate, hope over despair, courage over fear."

The helicopter completed its patrol and disappeared over the western ridge, but Eva knew it would be back. The Phoenix Order's surveillance was systematic, thorough, and designed to detect any approach to their fortress.

"When do we attempt the infiltration?" she asked.

"After dark. The thermal imaging will be more effective, but the acoustic and seismic sensors will be less sensitive. If we move carefully, if we avoid triggering motion detectors, we might reach the aqueduct entrance undetected."

Eva calculated the timeline in her head. "Sunset is at 16:30. Full darkness by 17:00. If we start the traverse at 18:00, we could reach the basement by 21:00."

"Giving us approximately twelve hours to complete the mission before the Christmas Eve deadline."

"Assuming everything goes perfectly. Assuming we find Klaus's research immediately. Assuming the synthesis process works flawlessly. Assuming we can deploy the counter-virus without complications."

"Assuming we survive the encounter with your father," Ben added quietly.

Eva felt tears stinging her eyes as the reality of their situation crystallized. They were about to attempt the most dangerous infiltration in human history, carrying the weight of four billion lives on their shoulders, facing an enemy who knew them better than they knew themselves.

"Ben," she said, her voice barely audible above the wind, "I need to tell you something."

"What?"

"If something happens to me in there, if Klaus captures me or if I'm killed, you have to promise me you'll continue the mission. You'll find a way to synthesize and deploy the counter-virus."

"Eva, I don't know enough about biochemistry to."

"You'll figure it out. You're brilliant and resourceful, and you have access to my mother's research. But more importantly, you understand what we're fighting for." Eva's eyes met his, and Ben saw the absolute determination there. "Promise me that my death won't stop you from saving those four billion lives."

Ben was quiet for a long moment, his hands covering hers in the snow. "I promise. But I need you to promise me something, too."

"What?"

"That you'll remember what your mother taught you. That love is stronger than hate. That hope is more powerful than fear. That the genetic code hidden in her lullaby isn't just a scientific formula, it's a message about the fundamental goodness of human nature."

"I promise."

They held each other on the snowy ridge, two people about to attempt the impossible, knowing that their love might be all that stood between humanity and extinction. Below them, the monastery waited, its secrets and dangers, its memories, and horrors.

Klaus would be preparing for their arrival, setting traps both physical and psychological, using every advantage of terrain and psychology to ensure their defeat. But Eva and Ben had something Klaus couldn't account for in his calculations.

They had witnessed human kindness in its purest form. They had experienced love in its most powerful expression. They had learned that the choice to be good was stronger than any genetic programming.

The monastery was both a fortress and a prison. Klaus had trapped himself inside his own ideology, surrounded himself with people who shared his hatred, and isolated himself from the very humanity he claimed to understand.

Eva and Ben were walking into that prison voluntarily. Still, they were bringing something with them that no fortress could contain.

They were bringing hope.

The sun was setting behind the western peaks, painting the snow-covered mountains in shades of gold and crimson. In three hours, they would begin their descent into the aqueduct system, starting a journey that would determine the future of human civilization.

"Eva," Ben said as they prepared to move to their infiltration position, "whatever happens in there, whatever Klaus says or does, I want you to remember something."

"What?"

"This morning, in Sankt Georgen, we saw humanity at its best. Johann and Margarete, their nephew, and the simple kindness of strangers helping strangers. That's what we're fighting for. That's what Klaus wants to destroy. That's why we can't let him win."

Eva nodded, feeling her resolve strengthen. Klaus might control the monastery, but he couldn't control the human capacity for love. The Phoenix Order might have superior numbers and technology, but they couldn't manufacture the kind of courage that came from fighting for something bigger than yourself.

"Fourteen hours until Christmas Eve," she said, checking her watch.

"Fourteen hours to save the world," Ben replied.

They began their careful descent toward the aqueduct entrance, moving with the patience of people who understood that humanity's future depended on their success. Behind them, the monastery continued its electronic vigil, sensors scanning, cameras recording, guards maintaining their posts.

But somewhere in that fortress, hidden in a laboratory that had once belonged to a brilliant woman who loved her daughter enough to encode salvation in a lullaby, lay the key to stopping the Phoenix Order's genocidal plan.

The mountain fortress was impregnable.

But love had always been the force that moved mountains.

The final infiltration was about to begin.

And with it, the last battle for humanity's soul.

Klaus thought he had prepared for every possibility.

He was about to discover that some things couldn't be defended against.

Some things were stronger than stone walls and armed guards.

Some things could penetrate any fortress.

Love was one of those things.

Hope was another.

And together, they were unstoppable.

The countdown continued.

Fourteen hours remained.

The end game had begun.

Time to discover if two people with nothing left to lose could save the world.

Time to find out if a daughter's love could overcome a father's hate.

Time to learn whether humanity deserved the future it was about to receive.

The mountain fortress waited.

But so did the power of love.

And in the end, love always wins.

Even when it seemed impossible.

Especially when it seemed impossible.

The infiltration was about to begin.

The final battle for humanity's future was at hand.

# Chapter 22: The Blizzard Infiltration

The blizzard arrived like a gift from God, or perhaps from Eva's mother, watching over them from whatever place the dead observed the living.

At 18:47, just as the last traces of twilight faded behind the Alpine peaks, the wind began howling through the mountains with supernatural fury. Within minutes, snow was falling so heavily that visibility dropped to mere meters. The sophisticated thermal imaging cameras that ringed the monastery's perimeter became useless, their sensors overwhelmed by the swirling chaos of ice and wind.

"Now," Ben whispered, his voice barely audible above the storm's roar.

They descended from their observation post like ghosts, using the blizzard's concealment to approach the entrance of the ancient Roman aqueduct. Eva's childhood memories guided them through terrain that would have been impossible to navigate without intimate knowledge of every rock, every tree, every subtle variation in the mountainside's contours.

The aqueduct entrance was exactly where Eva remembered it: a narrow opening hidden behind a screen of winter-bare bushes, disguised by centuries of weathering that made it appear to be nothing more than a natural cave. The Roman engineers who'd built this system two thousand years ago had never imagined their work would someday serve as an infiltration route for desperate scientists attempting to prevent global genocide.

"Are you certain this connects to the monastery?" Ben asked, studying the dark opening with the skeptical eye of someone whose life depended on accurate intelligence.

"I crawled through this entire system when I was eight years old," Eva replied, her breath forming clouds of vapor that were instantly torn away by the wind. "My mother was working late in her laboratory, Klaus was traveling, and I was bored enough to explore places that were supposed to be off-limits."

"And Klaus never discovered your explorations?"

"Klaus was already beginning to lose interest in family life by then. As long as I didn't interfere with his research, he didn't much care what I did with my time."

Ben nodded, understanding the psychology of a man whose obsessions had gradually consumed every human connection. "Lead the way."

The aqueduct's interior was a tunnel carved from living rock, wide enough for a single person to crawl through but too narrow for comfortable movement. The ancient stones were slick with moisture and ice, and the air carried the mineral scent of deep earth and flowing water.

Eva took point, her more petite frame allowing her to navigate the passage more easily than Ben's broader shoulders. Behind her, she could hear his controlled breathing, the slight scrape of equipment against stone, the careful placement of his hands and knees on surfaces that had been worn smooth by two millennia of water flow.

"How far to the first junction?" Ben asked, his voice echoing strangely in the confined space.

"Maybe two hundred meters. The passage splits there, with one branch leading to the old wine cellars and another connecting to the building's main water supply."

"Which route do we take?"

"Water supply. It connects directly to the basement levels where my mother's laboratory is located." Eva paused her forward movement, listening carefully to the sounds around them. "Ben, do you hear that?"

"What?"

"Water. Running water. There shouldn't be any active flow in this section during winter."

Ben listened more carefully, his trained ear picking up the subtle sound of liquid moving through stone channels. "Could be melting snow from the storm."

"Maybe. Or Klaus might have reactivated parts of the system for security purposes." Eva's scientific mind processed the implications. "If he's pumping water through the tunnels, he might be monitoring flow rates, pressure variations, anything that might indicate unauthorized access."

"Can we proceed without triggering sensors?"

"We'll find out." Eva's honesty was brutal but necessary. "The alternative is turning back, and we don't have time for alternative routes."

They continued through the ancient passage; their progress measured in meters rather than minutes. The Roman construction was masterful, but two thousand years of geological settling had created obstacles that required careful navigation. Fallen stones blocked some sections, forcing them to squeeze through gaps barely wide enough for human passage. Ice formations hung from the ceiling like crystalline daggers, threatening to break loose and alert any acoustic monitoring systems.

The first junction appeared exactly where Eva remembered it, marked by carved Roman numerals that had survived the centuries with remarkable clarity. The sound of the water was stronger here, echoing from multiple directions through the tunnel system.

"This way," Eva whispered, choosing the passage that led toward the monastery's foundation.

The tunnel began to slope upward, following the natural contours of the mountainside as it approached the building's basement levels. Eva could feel the weight of stone above them, millions of tons of medieval architecture pressing down on the ancient Roman engineering that carried them toward their destination.

"Movement," Ben said suddenly, his voice sharp with warning.

Eva froze, listening carefully. At first, she heard only the sound of water and wind, but then she detected what had alerted Ben: footsteps. They moved through the tunnels ahead of them, measured and regular.

"Guards," she breathed.

"How is that possible? I thought you said Klaus didn't know about this system."

"He shouldn't know about it. Unless..." Eva's mind raced through possibilities. "Unless someone has been exploring the facility's infrastructure more thoroughly than I anticipated."

The footsteps were getting closer, accompanied by the beam of a powerful flashlight that sent shadows dancing through the tunnel system. Ben motioned for silence, drawing the knife from his equipment belt with movements so careful they made no sound whatsoever.

A Phoenix Order operative appeared at the tunnel junction, his automatic weapon sweeping the passage with professional thoroughness. He was young, competent, and obviously following a specific patrol protocol that suggested this route was regularly monitored.

Ben moved like a predator, his CIA training allowing him to close the distance without triggering the guard's peripheral awareness. The combat was swift, silent, and absolutely lethal. Within seconds, the operative was dead, his neck broken with surgical precision.

"They know about the aqueduct," Ben said, searching for the dead man's equipment. "This isn't random exploration. This is a systematic patrol of a known infiltration route."

Eva felt her stomach clenched with dread. "Klaus anticipated this approach. He's been waiting for us to use my childhood knowledge against him."

"Which means there will be more guards. More patrols. More obstacles between us and your mother's laboratory."

Ben found what he was looking for in the guard's equipment: a radio handset and a detailed map of the tunnel system. The map was annotated with patrol routes, checkpoint schedules, and sensor locations that hadn't existed during Eva's childhood explorations.

"Look at this," Ben showed her the map. "They've installed motion detectors at every major junction. Acoustic sensors in the narrowest passages. Even pressure plates in sections where the tunnel floor has been reinforced."

Eva studied the security modifications with growing horror. Klaus hadn't just fortified the monastery's exterior. He'd turned the entire underground system into a maze of electronic surveillance and armed guards.

"How long until this patrol is missed?" she asked.

Ben checked the dead guard's watch and radio schedule. "Next check-in is at 20:30. That gives us maybe forty minutes before they realize something's wrong."

"Can we reach the basement levels in forty minutes?"

"Not through this route. The security is too heavy, the obstacles too numerous." Ben studied the map more carefully, his tactical mind working through alternatives. "But there's another possibility."

"What?"

"The water supply tunnel. According to this map, it's less heavily guarded because they consider it impassable during winter conditions."

"It might actually be impassable. That tunnel fills with runoff during snow melts, and with this blizzard..." Eva's voice trailed off as she calculated the risks.

"How dangerous?"

"Potentially fatal. Fast-moving water in a confined space, ice formations that could block exit routes, hypothermia from prolonged exposure to freezing temperatures."

Ben weighed their options with the cold logic of someone who'd spent years making life-or-death decisions. "What are our chances of survival through the water tunnel?"

"Maybe sixty percent if conditions are favorable."

"And our chances of fighting through the heavily guarded route?"

"Less than ten percent."

"Then we take the water tunnel." Ben's decision was immediate and final. "Better to risk drowning than guarantee capture."

They backtracked to the junction and entered the passage Eva had avoided, immediately understanding why the Phoenix Order considered it impassable. The tunnel was partially flooded with rushing meltwater that had turned the ancient stone channel into a subterranean river.

The water was shockingly cold, reaching their waists within the first few meters. Eva gasped as the icy flow penetrated her clothing, numbing her legs and making movement a struggle against the current's relentless force.

"Stay close," Ben shouted over the water's roar. "If you lose your footing, grab onto me immediately."

They waded through the tunnel like explorers in an underground nightmare, fighting the current while trying to maintain their footing on stones made treacherous by centuries of water erosion. The ceiling lowered as they progressed, forcing them to duck beneath formations that threatened to knock them unconscious.

"There," Eva pointed to a subtle change in the tunnel's construction ahead. "That's where the Roman work connects to the monastery's foundation. We're almost through the worst section."

But as they approached the connection point, Eva realized their problems were just beginning. The junction was fitted with a sophisticated filtration system that hadn't existed during her childhood, complete with electronic sensors that would detect any disruption in water flow.

"Motion detectors," Ben observed, studying the installation. "Pressure sensors. Flow monitors. Klaus has turned even the water supply into a security checkpoint."

"Can we bypass them?"

"Not without triggering alarms." Ben's assessment was grimly professional. "These systems are designed to detect any foreign object in the water flow, including human bodies."

Eva studied the sensor array, her scientific mind working through the challenge. The devices were clearly sophisticated, but they were also operating in an environment that presented unique opportunities for deception.

"Ben, do you still have explosives from your equipment kit?"

"Small charges for emergency breaching. Why?"

"Because I think we can create enough chaos to mask our passage." Eva's voice gained strength as the plan crystallized. "If we can trigger a controlled collapse in the tunnel section behind us, the resulting debris flow might overwhelm the sensors' ability to distinguish between rocks and human bodies."

Ben studied the tunnel construction, his tactical mind evaluating the feasibility. "The blast would have to be precisely timed. It is too small and it won't create enough debris. Too large and it might bring down the entire section."

"Can you, do it?"

"Maybe. But once we trigger the explosion, we'll have maybe five minutes before security teams respond to investigate the disturbance."

"Five minutes to traverse the sensors, reach the basement levels, and access my mother's laboratory." Eva's scientific training demanded honest assessment. "It's possible, but barely."

Ben was already placing the small explosive charges, his hands working with the precision of someone who'd done this countless times in hostile territory. "When I trigger the blast, you move immediately toward the sensors. Don't hesitate, don't look back, don't worry about anything except reaching the laboratory."

"What about you?"

"I'll be right behind you." Ben's smile was grim but genuine. "Eva, whatever happens in the next few minutes, remember that we've already accomplished something remarkable. We've proven that love is stronger than hate, that hope can triumph over despair."

The explosion was perfectly calibrated, sending a cascade of stone and debris rushing through the tunnel behind them. The noise was tremendous, echoing through the entire underground system with the force of an earthquake.

Eva moved instantly, diving through the swirling water and debris toward the sensor array. The filtration system's alarms were screaming, its detection algorithms overwhelmed by the chaos of rocks and ice flowing through the previously controlled environment.

She passed through the sensor field like a ghost, her body masked by the debris flow that Ben's precisely placed charges had created. Behind her, she could hear him following, his larger frame struggling against the current but maintaining progress through sheer determination.

They emerged from the water tunnel into the monastery's basement levels, soaked, exhausted, and hypothermic, but alive. The familiar stone corridors of Eva's childhood stretched ahead of them, but they had been transformed by Klaus's security modifications.

Motion sensors tracked their every movement. Automated weapon systems emerged from concealed positions in the walls. Blast doors that hadn't existed during her youth now divided the corridors into kill zones that could be sealed at Klaus's command.

"Welcome home, Eva," a voice said through speakers hidden in the ancient stonework.

Klaus. Her father's voice was tinged with the satisfaction of a chess master whose trap had finally been sprung.

"I've been expecting you," the voice continued. "Both of you. I must admit, the aqueduct approach was clever, though hardly unexpected. Did you really think I wouldn't investigate every route you might use to return to your childhood home?"

Eva felt the familiar chill of Klaus's psychological manipulation, but she also felt something else. Strength. The memory of Johann and Margarete's kindness, the warmth of Ben's love, and the knowledge that she carried her mother's genetic legacy encoded in childhood memories.

"Hello, Papa," she said, her voice steady despite the circumstances. "We need to talk."

"Indeed, we do, Meine Tochter. Indeed, we do."

The basement corridors stretched ahead of them, each step bringing them closer to the confrontation that would determine humanity's future. Klaus was waiting, prepared, confident in his advantages of terrain and psychology.

But Eva and Ben had something Klaus couldn't account for in his calculations.

They had each other. They had love. And they had the determination to honor Ingrid Richter's sacrifice by completing what she had started.

The infiltration was complete.

The final battle was about to begin.

And somewhere in the monastery's depths, a mother's love waited to save the world through a lullaby that held the secret of life itself.

Klaus thought he had prepared for every possibility.

He was about to discover that some things couldn't be defended against.

Love was one of those things.

Hope was another.

And together, they were unstoppable.

Ten hours remained until Christmas Eve.

Time to discover if two people with nothing left to lose could save four billion lives.

The blizzard infiltration was complete.

The real war was about to begin.

And in the end, love would triumph.

Even in the depths of a fortress built on hate.

Especially in the depths of a fortress built on hate.

The final confrontation awaited.

Father and daughter, face to face, with the future of humanity hanging in the balance.

The countdown continued.

The endgame had arrived.

# Chapter 23: The Gauntlet

T he corridor that had once led to Eva's favorite childhood hiding place was now a killing field designed by someone who knew her intimately.

Motion sensors tracked their every step as they moved through the monastery's basement levels. At the same time, automated weapon systems emerged from walls that had been solid stone during Eva's youth. The ancient architecture that had sheltered generations of monks had been perverted into a high-tech labyrinth where every familiar landmark concealed mortal danger.

"Left at the next junction," Eva whispered, her voice barely audible above the hum of surveillance equipment. "The passage leads to the main laboratory complex."

Ben nodded, his weapon drawn and ready, his eyes scanning for threats that could emerge from any direction. Behind them, alarms were beginning to sound as security teams responded to the explosion in the aqueduct system. They had perhaps five minutes before the monastery's defenders organized a coordinated response.

The junction Eva remembered as a simple intersection of stone corridors had been transformed into a checkpoint bristling with sensors and defensive positions. Blast doors could seal off any route within seconds, while concealed weapon emplacements covered every approach angle.

"This wasn't here when I was a child," Eva said, studying the modifications with growing unease. "Klaus has rebuilt the entire basement level."

"Not rebuilt," Ben observed, his tactical eye cataloging the defensive improvements. "Enhanced. He's used your childhood knowledge as a template, then designed countermeasures for every route you might remember."

The realization was chilling. Klaus hadn't just fortified the monastery against random intruders. He'd studied his daughter's behavior patterns, her childhood movements, her psychological tendencies, and created a killing field specifically designed to exploit her intimate knowledge of the facility.

"He's not just defending against us," Eva said quietly. "He's using my own memories as weapons."

The first security team appeared at the far end of the corridor, their movements coordinated and professional. Phoenix Order operatives are equipped with automatic weapons and full tactical gear. They advanced with the confidence of people fighting on familiar ground against known opponents.

"Down," Ben commanded, pulling Eva behind a stone buttress as bullets began chipping away at the ancient masonry.

The firefight was brief but intense. Ben's training enabled him to place precise shots that exploited gaps in the operatives' body armor, while Eva utilized her knowledge of the facility's layout to predict their movement patterns.

"Ventilation shaft," she pointed to a grating high on the wall. "It connects to the laboratory levels, but the passage is narrow."

"Can you fit through it?"

"Yes, but you'll have trouble. And if they seal the shaft while we're inside..."

"We'll be trapped." Ben finished the thought while calculating their remaining ammunition. Three magazines left, plus whatever they could scavenge from the dead operatives. Not enough for a prolonged engagement.

More footsteps echoed from multiple directions as additional security teams converged on their position. Klaus was deploying his forces with the precision of someone who'd planned this confrontation in meticulous detail.

"The shaft," Ben decided. "It's our best option."

Eva boosted herself up to the ventilation grating, her fingers working quickly to remove the screws that held it in place. The passage beyond was exactly as she remembered from childhood explorations, though motion sensors had been installed at regular intervals.

"They're monitoring the shafts too," she reported.

"Then we move fast and hope the alarms create enough confusion to mask our signatures."

They crawled through the ventilation system like desperate animals, the metal ductwork barely wide enough to accommodate Ben's broader frame. Behind them, Eva could hear security teams entering the corridor, their voices coordinating search patterns that would systematically eliminate every hiding place.

The shaft opened into the main laboratory complex, and Eva's breath caught as she saw what Klaus had built in the years since her escape. The facility that had once housed maybe twenty researchers now contained industrial-scale production equipment. These synthesis arrays could manufacture biological agents in quantities measured in tons rather than grams.

"My God," she whispered, taking in the scope of the operation. "This isn't just research anymore. This is weapons production."

Ben dropped from the ventilation shaft and immediately took cover behind a bank of synthesis equipment. "How many people would it take to operate a facility this size?"

"Hundreds. Maybe a thousand." Eva's scientific mind cataloged the equipment with growing horror. "Ben, this isn't just about the Lethe Virus. They're producing multiple biological agents, different targeting parameters, and various deployment mechanisms."

She moved through the laboratory with the confidence of someone who understood the science behind the terror. Protein synthesis arrays, genetic modification chambers, and viral cultivation systems operating at scales that dwarfed anything she'd imagined during her unwitting participation in the project.

"Here," Eva pointed to a bank of computer terminals displaying production schedules. "Look at this."

The screens showed a manufacturing timeline that made the Christmas Eve attack seem like a preliminary test rather than the main event. Dozens of different biological agents, each targeting specific populations, are designed for deployment through various vectors.

"Phase One: Lethe Virus, airborne deployment, forty-seven release points," Eva read from the schedule. "Phase Two: Prometheus Strain, water supply contamination, one hundred twelve target cities. Phase Three: Nemesis Pathogen, food supply infiltration, global deployment."

Ben studied the timeline with growing dread. "They're not planning a single attack. They're planning a systematic campaign of biological warfare that could last years."

"Each phase is designed to eliminate different population groups based on genetic markers, geographical location, and dietary patterns." Eva's voice carried the hollow shock of someone confronting the full scope of genocidal planning. "Klaus isn't just trying to kill four billion people. He's trying to reshape human evolution through selective biological warfare."

The laboratory's intercom system crackled to life, Klaus's voice filling the space with paternal disappointment.

"Eva, I'm impressed by your infiltration skills, but you're seeing only a fraction of what we've accomplished here. The Phoenix Order's vision extends far beyond the limited scope you've imagined."

Eva felt the familiar chill of her father's psychological manipulation, but she forced herself to respond. "It's not a vision, Papa. It's genocide disguised as evolution."

"It's salvation disguised as science," Klaus replied. "Humanity is dying, Eva. Genetic degradation, overpopulation, and the inevitable collapse of civilized society. We're offering the species a chance to survive, to evolve, to become something better than what we are."

Ben moved through the laboratory while Klaus spoke, his tactical mind cataloging defensive positions and potential weapons. The facility was a maze of sophisticated equipment. Still, it was also a target-rich environment where a few well-placed shots could cause catastrophic damage.

"Show yourself, Papa," Eva called out. "If you believe so strongly in your cause, have the courage to face me directly."

"Courage?" Klaus's laugh carried a note of genuine amusement. "My dear daughter, I've dared to sacrifice everything for humanity's future. My family, my conscience, my peace of mind. What have you sacrificed beyond your comfortable illusions about human nature?"

Eva felt tears stinging her eyes, but her voice remained steady. "I've sacrificed my faith in you. I've given up the illusion that the man who raised me was anything more than a monster in a father's clothing."

The words hung in the air for a moment before Klaus responded, his voice carrying a note of what might have been pain.

"Then you understand the price of evolution. The necessity of becoming something harder, colder, more capable of making the choices that ensure survival."

"I understand that you've lost your humanity in pursuit of an ideology that would destroy everything worth preserving about human nature."

Ben had reached the far end of the laboratory, where a sealed door marked with biohazard warnings led to what appeared to be the virus storage facility. The lock required dual authorization, exactly as Eva had described from her father's files.

"Eva," Ben called softly. "I've found the storage vault."

She joined him at the door, studying the security system with professional interest. "Biometric scanners for Klaus and Wilhelm von Hess. Voice recognition backup. Probably thermal imaging to ensure the authorized users are alive and present."

"Can you bypass it?"

"Maybe. If Klaus still trusts me enough to maintain my emergency access codes." Eva placed her hand on the scanner, holding her breath as the system processed her biometric data.

The scanner flashed red. Access denied.

"He's revoked my clearance," Eva said, disappointment clear in her voice.

"Then we need to find another way." Ben studied the door's construction, noting the reinforced steel frame and electromagnetic locks. "Explosives?"

"Too dangerous. If we damage the storage containers, we could release biological agents in concentrations that would kill us instantly."

The sound of approaching footsteps made them both freeze. Security teams were systematically searching the laboratory complex; their movements coordinated through radio communications that suggested military-level training.

"This way," Eva led Ben toward the rear of the facility, where her mother's private laboratory had been located. "If Klaus maintained her workspace, there might be equipment we can use."

They moved through the laboratory like ghosts, using the complex equipment arrays as cover while security teams conducted their search. Eva's knowledge of the facility's layout allowed them to stay one step ahead of their pursuers, but she could feel the net tightening with each passing minute.

Her mother's laboratory was exactly as Eva remembered it, preserved like a shrine to the brilliant woman who had once worked there. The equipment was older than the industrial-scale machinery Klaus had installed. Still, it was also more sophisticated, designed for precision rather than mass production.

"This is where she developed the counter-virus," Eva said, her voice soft with memory. "Where she encoded the genetic sequence in my lullaby."

Ben studied the laboratory equipment while Eva accessed her mother's computer terminal. "Can you synthesize the counter-virus here?"

"Maybe. The equipment is sophisticated enough, and I have the genetic sequence from the lullaby." Eva's fingers flew across the keyboard as she pulled up her mother's research files. "But I need samples of the current virus to verify compatibility."

"Which means we still need access to the storage vault."

"Or we need Klaus to provide the samples himself." Eva's voice carried a note of grim determination. "Ben, I think I know how to force him into a direct confrontation."

Before Ben could ask what, she meant, Eva had activated the laboratory's intercom system and was broadcasting throughout the facility.

"Attention all Phoenix Order personnel," her voice echoed through the monastery's corridors. "This is Dr. Eva Richter. I'm currently in possession of my mother's complete research files, including the genetic codes for biological agents that could neutralize your entire operation."

The response was immediate. Alarms began blaring throughout the facility as security teams redirected toward the laboratory complex. But Eva continued her broadcast, her voice steady despite the chaos she was creating.

"Klaus, you have exactly ten minutes to meet me in Mother's laboratory for a direct negotiation. Alone. Unarmed. If you refuse, I'll release her research files to every major news organization, every government health agency, every international court in the world."

Ben stared at her in amazement. "Eva, you're bluffing. You don't have those files."

"I know. But Klaus doesn't know that. And his ego won't allow him to ignore a direct challenge from his daughter."

The intercom crackled with Klaus's response, his voice carrying cold fury.

"You're playing a dangerous game, Eva. One that could cost you everything."

"I've already lost everything, Papa. My mother, my family, my faith in human decency. What's left for you to take?"

"Your life," Klaus replied simply. "And the life of the man you've chosen to love."

Eva felt Ben's hand close over hers, providing strength and comfort in the face of her father's threats. "Then come and take them. But come alone, or I'll make sure the world knows exactly what the Phoenix Order is planning."

The silence that followed was pregnant with tension. Eva could hear security teams repositioning outside the laboratory, their movements suggesting they were establishing a perimeter rather than preparing for immediate assault.

Klaus was considering her challenge, weighing the risks and benefits with the cold calculation that had made him one of the world's most dangerous men.

"Very well," he said finally. "Ten minutes. Mother's laboratory. Just you and me, Eva. As it should have been from the beginning."

The intercom went silent, leaving Eva and Ben alone in the laboratory where Ingrid Richter had once worked to save the world from her husband's madness.

"He's coming," Eva said unnecessarily.

"Are you ready for this?"

Eva looked around the laboratory, taking in her mother's equipment, her research notes, and the place where love had battled hate and lost. But she also saw something else. She saw the place where love might finally triumph, where a daughter's courage might complete what a mother's sacrifice had begun.

"I'm ready," she said, her voice steady with absolute conviction.

The gauntlet was ending.

The final confrontation was about to begin.

And in a laboratory dedicated to preserving life, the future of humanity would be decided by a conversation between a father and daughter who had chosen opposite sides in the war for the human soul.

Nine hours remained until Christmas Eve.

Nine hours to save the world.

Nine hours to prove that love was stronger than hate, hope more powerful than despair, courage greater than fear.

The countdown continued.

The endgame had arrived.

Klaus was coming.

And Eva was ready to face the monster her father had become.

Time to discover if a daughter's love could overcome a father's ideology.

Time to learn whether humanity deserved the future it was about to receive.

The final battle for the human soul was about to begin.

In her mother's laboratory.

Where it had always been meant to end.

Where love would finally have its chance to triumph over hate.

The gauntlet was complete.

The real war was about to begin.

And this time, love would not lose.

This time, hope would triumph.

This time, the future would belong to those who chose compassion over cruelty.

Klaus thought he had won.

He was about to discover he had lost long ago.

The moment he chose ideology over love.

The moment he became a monster.

The final confrontation awaited.

Father and daughter.

Love and hate.

The future of humanity hanging in the balance.

Nine hours.
And counting.

# Chapter 24: The Laboratory of Horrors

The security door to the main laboratory complex required both Eva's biometric scan and her mother's old access code, a combination that shouldn't have worked after fifteen years. Yet the heavy steel barriers parted with a pneumatic hiss, revealing a corridor that stretched into antiseptic whiteness.

"They wanted us to get this far," Ben murmured, his MP5 tracking the empty hallway. Shell casings from their running firefight through the upper levels littered the floor behind them, and his tactical vest showed scorch marks from near misses. "This feels like a trap."

Eva knew he was right, but they had no choice but to do so. The virus was set to release globally in less than thirty-six hours. Whatever horror awaited them in her father's laboratory, they had to face them.

The corridor's walls were lined with observation windows, each revealing a different research chamber. Eva's breath caught as she recognized the layout. She'd helped design these containment systems, believing they would house beneficial genetic research. Now, approaching the first window, she saw what they truly contained.

"Dear God," she whispered.

Behind the reinforced glass, a test subject lay strapped to a medical table. Male, perhaps thirty years old, his body showed the ravages of viral exposure. But this wasn't the clean, engineered death she'd expected. The man's skin had developed a mottled, scaly texture. His limbs had elongated unnaturally, fingers stretching into claw-like appendages. He was still alive, eyes rolling wildly as monitors tracked his vital signs.

"Variant Seven," Eva read from the digital display. "Rapid evolutionary mutation triggered by specific genetic markers found in sub-Saharan African populations. Subject has survived forty-three hours post-exposure."

Ben moved to the next window, his face paling. "This one's worse."

Eva joined him and immediately wished she hadn't. The female subject in this chamber had been exposed to a different variant. Her body was literally consuming itself, flesh dissolving and regenerating in a horrific cycle. According to the monitors, she'd been in this state for six days.

"Southeast Asian genetic markers," Eva noted, her scientific mind processing even as her humanity recoiled. "The virus identifies specific hereditary sequences and triggers different responses. This isn't just about killing. It's about making examples."

"Psychological warfare," Ben agreed grimly. "Release these variants in targeted populations, and the terror alone would cause mass panic. Governments would surrender just to make it stop."

They moved deeper into the laboratory complex, each observation window revealing new horrors. A chamber filled with children's toys sat empty, its purpose too terrible to contemplate. Another held what appeared to be a family group, all in various stages of transformation.

"The Kowalski family," Eva read from a research tablet she left carelessly on a desk. "Polish immigrants, third generation. Genetic markers indicate Slavic ancestry with minor Jewish heritage. Exposed to Variant Twelve to test genealogical detection capabilities."

Her hands shook as she scrolled through the data. The virus could detect genetic heritage going back centuries, identifying bloodlines that the Phoenix Order deemed unworthy. The sophistication was beyond anything she'd imagined possible.

"Eva." Ben's voice carried a warning. He'd moved ahead to a set of double doors marked with biohazard symbols. "You need to see this."

The central laboratory sprawled before them like a cathedral of science perverted to monstrous ends. Dozens of workstations lined the walls, each dedicated to a different viral variant. Holographic displays showed genetic sequences of breathtaking complexity, the building blocks of selective genocide.

But it was the center of the room that made Eva's knees weak.

A massive holographic globe rotated slowly, marked with release points for the virus. Not just twelve cities, as they'd believed, but hundreds of locations. Each point pulsed with a different color, indicating which viral variant would be deployed in that location.

"They're not just targeting cities," Ben observed, studying the display. "Rural areas, refugee camps, indigenous reservations. Anywhere they've identified genetic populations for elimination."

Eva approached the central console, her fingers flying across the interface. Her access codes still worked, probably by design. Klaus wanted her to see this, to understand the full scope of what she'd helped create.

"Two hundred and seventeen release points," she confirmed, her voice hollow. "Each with a specific viral variant designed for the local population's genetic markers. The initial releases in Lagos, Kinshasa, and Cairo were just tests. This..." she gestured at the globe, "this is extinction-level planning."

"Can you identify which variants go where?" Ben asked, photographing everything with a secured phone.

Eva pulled up the deployment matrix, and her heart sank further. "It's precisely targeted. Variant One through Fifteen target what they call 'primitive genetic lines'—essentially anyone with significant African, indigenous American, or Australian Aboriginal ancestry. Variants Sixteen through Thirty focus on 'mongrel populations,' their term for mixed heritage individuals."

"Let me guess," Ben said darkly. "Northern Europeans are largely spared."

"Not entirely." Eva found herself laughing bitterly. "Variants Thirty-One through Thirty-Five target what they consider 'corrupted Aryan lines.' Europeans with any Jewish, Roma, or Slavic ancestry beyond acceptable parameters. My father was nothing if not thorough."

A new horror occurred to her. She quickly pulled up her own genetic profile from the database. There it was, flagged in green: 'Acceptable breeding stock. Preserve for New Eden Protocol.'

"They've categorized everyone," she breathed. "Every researcher, every guard, every member of the Phoenix Order. Those marked green survive to build their new world. Everyone else..."

"Dies screaming," Ben finished, his jaw tight with anger.

Eva moved to another workstation; one she recognized as her father's personal terminal. Password protected, but she knew Klaus's arrogance. He'd use something meaningful to him. She tried her mother's maiden name. Nothing. Her own birthday. Failed.

Then, with a sick feeling, she typed the date of her mother's death.

The terminal unlocked.

"Of course," she muttered, accessing Klaus's private files. What she found there made the previous horrors pale in comparison.

"Project Prometheus - Final Phase," she read aloud. "Upon successful global deployment, survivor populations will be managed through ongoing genetic modification. The weak will be sterilized. The strength will be enhanced. Within three generations, Homo sapiens will be replaced by Homo superior."

"They're not just committing genocide," Ben said slowly. "They're trying to forcibly evolve the human species."

"It gets worse." Eva opened another file, this one containing video logs. Klaus's face appeared on screen, dated just a week ago.

"Personal log, December 17th," her father's recorded voice began. "Eva's work on the delivery system exceeded all expectations. The aerosolized particles can survive in atmospheric conditions for up to fourteen days, ensuring complete saturation of target areas. I'm so proud of her, even if she doesn't understand the gift she's given humanity."

Eva wanted to vomit. Her breakthrough in viral delivery, the work that had won her international acclaim, had been the final piece of the puzzle for global genocide.

The video continued: "Wilhelm insists we proceed on schedule, but I worry about Eva. Her emotional attachment to outdated moral frameworks could complicate things. Still, her genetic profile is too valuable to waste. After the cleansing, she'll understand. She'll see the beauty of a world freed from genetic impediments. Perhaps she'll even thank me."

"Delusional bastard," Ben growled.

Eva clicked to another file, this one labeled "Contingency Protocols." Her blood ran cold as she read.

"If primary deployment fails, secondary systems will activate automatically. Viral samples have been placed in water treatment facilities, food distribution centers, and transportation hubs worldwide. The Phoenix Order has members in every major government, every international organization. Even if we stop the primary release, the backup systems ensure victory."

"How do we fight something this comprehensive?" Ben asked, though his tone suggested he already knew the answer.

Eva frantically searched for weakness in the plan, some oversight they could exploit. That's when she found it, buried in a subsection of technical specifications.

"The variants," she said suddenly. "They're all based on the same core virus. My mother's original work, modified and weaponized. If we could access that base sequence, create a universal antibody..."

"You said the original was destroyed."

"The physical samples were. But the genetic sequence..." Eva's fingers flew across the keyboard, diving deep into archived data. "Every variant references the core structure. If I can reverse-engineer from the modifications, extract the common elements..."

An alarm pierced the laboratory's silence. Red lights began flashing as a computerized voice announced: "Lockdown protocol initiated. Containment breach in sectors seven through twelve."

"They're releasing the test subjects," Ben realized, checking his weapon. "Driving us forward."

Eva grabbed a portable drive and downloaded everything she could. The genetic sequences, deployment plans, personnel files—evidence of the Phoenix Order's crimes. "We need the synthesis equipment in the lower laboratory. If I can reconstruct the original virus, we might—"

Glass shattered in the observation windows. The test subjects weren't just being released; they were breaking free. The man with scaled skin crashed through the barrier, his mutated form moving with inhuman speed. Others followed, their virus-ravaged bodies driven by engineered aggression.

"Run!" Ben shoved Eva toward the far exit as he opened fire. The scaled man took multiple rounds before dropping, greenish blood spattering the pristine floor.

They sprinted through the laboratory as chaos erupted around them. Test subjects in various stages of mutation broke free from their chambers. Some attacked each other, viral programming turning them into weapons. Others focused on Eva and Ben, drawn by pheromone markers she'd unknowingly designed into the variants.

A woman whose skin had crystallized into sharp edges lunged at Eva. Ben intercepted, using his rifle stock to knock her aside, but not before one of her razor-like fingers sliced through his vest.

"Keep moving!" he shouted, blood seeping through the tear.

They reached the emergency stairwell just as Klaus's voice echoed through the laboratory speakers:

"Eva, my dear daughter. Do you see now? Do you understand the necessity of our work? These poor souls represent genetic dead ends, evolutionary failures. The virus simply accelerates what nature would do eventually. We're not monsters—we're gardeners, pruning humanity's tree so it can grow strong."

Eva slammed through the stairwell door, her father's words chasing them into darkness. Behind them, the laboratory had become a battleground for mutated humanity, a preview of the horror that awaited the world.

But she had the data now. The genetic sequences, the deployment plans, the core viral structure hidden within the variants. It wasn't much against the Phoenix Order's vast conspiracy, but it was a beginning.

As they descended toward the synthesis laboratory, Eva made a silent promise. She would stop this madness, would save everyone she could. And when it was over, she would ensure that her father and his Phoenix Order faced justice for their crimes against humanity.

The virus had been designed to create a new world through selective death.

Eva would use her mother's legacy to ensure that the world never came to be. Even if it cost her everything.

# Chapter 25: The Father's Trap

The central laboratory hummed with the steady pulse of refrigeration units and air filtration systems, a mechanical heartbeat that Eva had once found comforting. Now, as she and Ben crept through the shadows between towering equipment racks, every sound felt like a countdown to disaster.

"There," Eva whispered, pointing to the main terminal cluster. The holographic displays cast an eerie blue glow across the sterile white floor. "The complete viral database should be accessible from that station."

Ben moved with practiced silence despite his tactical gear, his MP5 sweeping each corner before they advanced. The laboratory stretched before them like a technological cathedral, its vaulted ceiling disappearing into darkness above. Automated systems continued their programmed tasks, oblivious to the intruders who had fought their way through three levels of security to reach this point.

Eva's fingers trembled as she approached the terminal. Her mother had designed this interface twenty years ago, back when the research had noble intentions. Back when they'd believed they were saving humanity, not engineering its selective destruction.

"How long do you need?" Ben asked, maintaining watch while she initiated the login sequence.

"Five minutes to access the files, another three to download the kill switch data." Her biometric scan flashed green—her father's arrogance in keeping her access active would be his downfall. "Then we can synthesize the counter-virus and—"

The laboratory's main lights blazed to life, flooding the space with harsh fluorescence. Eva's heart seized as a familiar voice echoed from hidden speakers.

"Hello, Eva."

Klaus Richter stepped from behind a cryo-storage unit, his white lab coat pristine, his expression a mask of paternal disappointment mixed with scientific curiosity. He looked exactly as he had during her childhood—distinguished, controlled, utterly sure of his righteousness.

Ben's weapon snapped toward the threat, but Klaus raised a hand, showing a small device. "I wouldn't, Mr. Carter. This dead man's switch is connected to the laboratory's emergency containment protocols. If my heart stops, this entire level floods with nerve gas. We'll all die together—poetic, wouldn't you say?"

"Klaus," Eva breathed, her voice catching on years of complicated love and fresh betrayal. "Please. You don't have to do this."

"Don't I?" Her father moved closer, his footsteps echoing in the vast space. "You know, I've been tracking you since you left. Your biometric signature is quite unique—a gift from your mother's genetic modifications during pregnancy. Every security camera, every sensor array, every Phoenix Order facility has been searching for my wayward daughter."

The terminal behind Eva chimed softly. Download initiated. Three minutes and forty-seven seconds remaining. She needed to keep him talking.

"The virus will kill millions," she said, forcing steadiness into her voice. "Innocent people. Children. How can you justify that?"

Klaus's laugh was soft, almost fond. "Innocent? There are no innocent people, Eva. Only those who contribute to human evolution and those who hold it back. The Phoenix Order isn't about destruction—it's about transformation. We're saving humanity from itself."

"By committing genocide?" Ben's voice cut through the father-daughter tension like a blade. "By targeting specific genetic markers? You're no different than the monsters who built the camps."

Klaus's eyes flashed with genuine anger. "How dare you compare us to those crude butchers? They acted from hate and ignorance. We act from love—love for humanity's potential. Every species must evolve or face extinction. We're simply... accelerating natural selection."

A metallic clang echoed through the laboratory as heavy security doors slammed shut. Red emergency lights began pulsing, and Eva's blood ran cold as she recognized the lockdown sequence.

"Did you really think I'd make it so easy?" Klaus asked, producing a tablet from his coat. "This facility has been my life's work. Every system, every protocol, every line of code—I know them all. You're not downloading the kill switch data, Eva. You're downloading exactly what I want you to download."

"No." Eva spun back to the terminal, her fingers flying across the holographic interface. But even as she tried to abort the transfer, she saw the truth in the data streams. The files were corrupted, infected with polymorphic code that would destroy any system that tried to synthesize the counter-virus.

"A father knows his daughter," Klaus continued, his voice taking on a lecturing tone she remembered from childhood. "I knew you'd come back. Your mother's weakness lives in you—that misguided belief that individual lives matter more than the species' survival. So I prepared a special welcome."

The download completed, and immediately, alarms shrieked through the laboratory. Warning messages flashed across every screen: VIRAL CONTAINMENT BREACH. EMERGENCY PROTOCOLS ACTIVATED.

"You're insane," Eva whispered, backing away from the terminal. "You've released the virus here? Now?"

"Not the Lethe strain," Klaus corrected with scientific precision. "Something special I developed just for this moment. A faster-acting variant. We have perhaps ten minutes before the aerosolized particles reach lethal concentration. Unless, of course, you'd like to input the proper shutdown code?"

Ben kept his weapon trained on Klaus while pulling Eva behind an equipment rack. "He's bluffing. No one's that crazy."

"You don't know my father," Eva said, her mind racing through possibilities. The laboratory's ventilation system had already reversed, sealing them in with whatever Klaus had released. "He's committed to his cause beyond reason."

"Beyond reason?" Klaus actually looked hurt. "Eva, everything I've done has been supremely logical. The human genome is stagnating. Our medical advances keep the weak alive, letting inferior genes propagate. In two generations, we'll be so genetically compromised that the species will collapse. The Phoenix Order offers controlled reduction—saving the strong, eliminating the weak. It's simple mathematics."

"People aren't numbers!" Eva shouted, surprising herself with the vehemence. "They're not data points in your sick experiment!"

"Aren't they?" Klaus touched his tablet, and holographic displays materialized around them, showing viral models and propagation scenarios. "Global population: eight billion. Sustainable population with current resources: two billion. The choice is simple—controlled selection now or chaotic collapse later. Which is more humane?"

A new alarm joined the cacophony—biohazard warnings flashing amber. Eva felt the first tickle in her throat, though she couldn't tell whether it was from the virus or fear.

"Where's the real kill switch data?" she demanded. "Mother discovered it, didn't she? That's why you murdered her."

Klaus's composed mask finally cracked. "I didn't murder her. The cancer—"

"Liar!" The word tore from Eva's throat. "I've seen the medical files. Induced cellular mutation. You poisoned her slowly, made it look natural. Because she threatened your perfect plan."

"She betrayed everything we'd worked for!" Klaus's control shattered completely, revealing the fanatic beneath the scientist. "Twenty years of research, and she wanted to destroy it all because she couldn't see the bigger picture. Just like you."

Ben moved subtly, trying to flank Klaus while keeping the dead man's switch in sight. But the older man noticed, raising the device higher. "Ah, the infamous Ben Carter. The analyst whose conscience cost forty-seven lives in Syria. Tell me, do those faces haunt you? Do you see them when you close your eyes?"

"Every night," Ben admitted quietly. "The difference between us is that I know I'm a killer. You've convinced yourself you're a savior."

"Because I am!" Klaus's eyes blazed with messianic fervor. "The Phoenix Order will save humanity from itself. Yes, billions will die. But those who survive will be stronger, smarter, and more adaptable. We're not destroying the human race—we're creating Humanity 2.0."

The air grew thick, and Eva tasted metal on her tongue. Whatever Klaus had released was working fast. She noticed Ben's breathing had become labored, though he maintained his tactical position.

"The kill switch," Eva pressed. "You said Mother discovered it. Where is it?"

Klaus smiled, the expression both proud and terrible. "You want to know the beautiful irony? There is no kill switch. Your mother believed she'd found one, a genetic sequence that could neutralize the virus. But I let her think that. Let her hope. Even encoded it in those silly lullabies she sang to you, knowing one day you'd come looking for it."

The words hit Eva like physical blows. "The lullaby... the memory... It's all fake?"

"Not fake. Just... incomplete. The sequence exists, but it only works with the original viral strain. The one we're releasing has evolved beyond that vulnerability. Your mother died believing she'd hidden the key to stopping us. Such tragic faith in a lie."

A crash echoed from somewhere deeper in the facility. Security forces, Eva realized. Her father's backup arrived to ensure their trap was complete.

"But I'm not without mercy," Klaus continued, producing a small silver vial from his pocket. "This contains the antidote to what I've released here. Enough for one person. You can save yourself, Eva. Join us. Your genetic profile marks you as worthy of survival. Help us build the new world."

"Never." The word came out as a growl.

"Then save him." Klaus gestured to Ben, who was now visibly struggling to breathe. "Your lover. Your protector. One dose. Choose quickly—the neural damage becomes irreversible in about three minutes."

Eva looked at Ben, saw the determination in his eyes even as his body betrayed him. He shook his head slightly, telling her without words not to take the deal. But how could she watch him die?

"There's a third option," she said quietly.

Before Klaus could react, Eva lunged forward, not for the antidote, but for the primary data terminal. Her fingers found the emergency purge command, and she input her mother's birthdate—the one password Klaus would never use.

"No!" Klaus dropped the dead man's switch and reached for her.

The laboratory's screens exploded with cascading delete commands. Twenty years of research, thousands of viral samples, the entire Phoenix Order database—all of it began erasing in a catastrophic chain reaction.

Klaus struck her across the face, sending her sprawling. "You stupid girl! You've doomed humanity!"

But Ben was already moving. The dead man's switch hadn't triggered—Klaus's arrogance had betrayed him with a bluff. Ben's weapon spoke three times, the shots grouped perfectly center mass.

Klaus staggered, looking down at the spreading crimson on his white coat with genuine surprise. "Eva..." he wheezed, reaching toward his daughter. "I only wanted... to make you... proud..."

He collapsed, the silver vial rolling from his fingers across the polished floor.

Eva scrambled for the antidote as the data purge continued around them. Only one dose. She looked at Ben, who had dropped to his knees, blood frothing at his lips from the viral damage to his lungs.

"Take it," he gasped. "The world... needs you..."

"The world needs us both." Eva cracked the vial open, pouring half its contents down Ben's throat before consuming the rest herself. It might not be enough for either of them, but she wouldn't choose between them. They'd survive together or die together.

The laboratory doors burst open as Phoenix Order security forces stormed in. But they stopped short, seeing Klaus dead and their life's work deleting itself from every server.

"Sir? What are your orders?" One soldier spoke into his radio. "The primary data core is compromised. Dr. Richter is down. Should we—"

The response crackled through: Wilhelm von Hess himself. "Seal the laboratory. Let them die with their principles. We proceed with the release as planned."

The security forces withdrew, and the massive doors sealed again. Eva cradled Ben's head in her lap as the partial antidote fought the virus in both their systems.

Around them, screens went dark one by one as twenty years of genocide research vanished into digital oblivion.

"Eva," Ben whispered, his breathing slightly easier. "The virus... still out ther e..."

"I know." She looked at her father's body, feeling grief and rage in equal measure. "But we're not dead yet. And mother's lullaby—even if it's incomplete, it's still a start. We just need to reach the synthesis lab."

Alarms continued their symphony of disaster as Eva helped Ben to his feet. They'd destroyed the primary research, but the Phoenix Order's plan remained in motion. Twelve cities. Millions of lives. And somewhere in her mother's hidden legacy lay the key to stopping it all.

If they could survive the next hour.

If they could escape the trap that had claimed her father's twisted soul.

If love and determination could triumph over calculated evil.

Eva looked into Ben's eyes, seeing her own resolve reflected there. They'd come too far to fail now. Together, they stumbled toward the laboratory's emergency exit, leaving Klaus Richter's body behind with his dreams of evolutionary supremacy.

The Phoenix Order had taken everything from her—her mother, her father, her life's work. But they'd made one crucial mistake.

They'd left her with something to fight for.

# Chapter 26: The Mother's Gift

The synthesis laboratory lay three floors below, but it might as well have been on another planet. Eva supported Ben's weight as they stumbled through the emergency stairwell, their footsteps echoing off concrete walls. The partial antidote had bought them time, nothing more. She could feel the virus still working in her system, a burning sensation that started in her lungs and spread outward like liquid fire.

"Wait," Ben gasped, leaning heavily against the wall. Blood spotted his lips despite the antidote. "Need... a moment..."

Eva checked her watch. Seventeen minutes since the laboratory lockdown. The Phoenix Order security forces would establish a perimeter, waiting for the virus to do its work. But something nagged at her—why had von Hess ordered them sealed in rather than simply executed?

"We have to keep moving," she urged, though her own legs trembled with exhaustion. "The synthesis lab has its own filtration system. If we can reach it—"

A metallic groan interrupted her. The building's ventilation system was reversed, creating negative pressure to contain the viral release. Standard protocol, one that her mother had designed. The irony tasted bitter as blood.

Ben pushed himself upright, soldier's discipline overcoming physical degradation. "Your father... said something about test sites. We need to know—"

"Later. First, we survive."

They descended another floor, and Eva's keycard still worked at the security checkpoint. Her father's oversight or deliberate trap? No way to know. The corridor beyond stretched empty and sterile, emergency lighting casting long shadows.

That's when she saw it.

Her mother's office, untouched after fifteen years. The nameplate still read "Dr. Elena Richter - Director of Bioethics." Eva had walked past it a hundred times during her years at the facility, never allowed inside. Klaus had declared it a shrine to her memory, off-limits to everyone.

"Eva, we don't have time," Ben warned, but she was already swiping her card.

The lock clicked open.

Inside, time had stopped. Her mother's desk remained exactly as she'd left it—papers neatly stacked, coffee mug waiting to be washed, a photo of young Eva building sandcastles on the Baltic coast. But it was the wall of filing cabinets that drew Eva's attention. Physical files in an age of digital storage. Her mother had been paranoid about electronic surveillance and insisted on hard copies for sensitive materials.

"Help me search," Eva said, yanking open drawers. "Anything about the original virus, the kill switch, Project Phoenix—"

Glass shattered in the corridor. They weren't alone anymore.

Ben limped to the door, checking his weapon. "Three magazines left. Whatever you're looking for, find it fast."

Eva's fingers flew through folders, decades of research flashing past. Viral cultures, genetic sequences, ethical reviews—and then she found it. A manila folder marked only with her birthday. Inside, her mother's handwriting, shaky and desperate:

*My darling Eva,*

*If you're reading this, then Klaus has won and I have failed. By the time you find these words, you'll know the truth—the cancer was a lie, a murder made to look like mercy.*

Eva's hands trembled as she continued reading.

*The Phoenix Order recruited your father when you were seven. They promised him unlimited resources, freedom from ethical constraints, a chance to reshape humanity itself. I was foolish enough to believe we could control it, use their funding for legitimate research. By the time I understood their true goals, we were trapped.*

Footsteps in the corridor. Ben fired twice, driving back unseen threats. "Eva!"

She read faster, her mother's words blurring through tears.

*The virus Klaus developed is beyond horror. It targets specific genetic markers, yes, but not randomly. The Phoenix Order has been collecting genetic data for decades, including ancestry services, medical records, and blood drives. They know exactly who will live and who will die. Entire populations marked for extinction with surgical precision.*

*I tried to stop it. Developed what I thought was a kill switch, a genetic sequence that would neutralize the virus. Klaus let me believe I'd succeeded, even helped encode it in the lullabies I sang to you. But it was always incomplete, missing crucial elements he kept hidden.*

A grenade rolled into the corridor. Ben grabbed Eva, pulling her behind the heavy desk as the explosion shook the room. Shrapnel peppered the walls where they'd been standing.

"We're pinned," Ben shouted over the ringing in their ears. He was right—only one exit, and Phoenix Order forces controlled it.

Eva clutched the letter, reading the final paragraphs:

*But I didn't stop fighting. In these filing cabinets, you'll find my real work. Not a kill switch—something better. A genetic mirror that turns the virus against itself. The Phoenix Order's weapon of selection becomes a tool of universal immunity. The strong don't survive while the weak die. Everyone survives, or no one does.*

*Klaus had me killed when he discovered this research. Made it look like cancer to avoid questions. He never found these files—he was too sentimental to clear out my office, too proud to believe I could have outsmarted him.*

Another explosion, closer this time. The office windows shattered, and Eva heard Wilhelm von Hess's voice over a loudspeaker:

"Dr. Richter, Mr. Carter. You've caused considerable inconvenience. Surrender now, and I promise your deaths will be painless. Continue resisting, and I'll ensure you watch each other die very slowly indeed."

Eva yanked open more filing cabinets, searching desperately for her mother's hidden research. Ben crouched by the doorway, conserving ammunition, but she could see the blood soaking through his tactical vest. The partial antidote was failing.

Then she found it, not in the filing cabinets, but hidden inside the coffee mug—a micro-drive disguised as ceramic glazing. Her mother's paranoia had reached genius levels.

"Got it!" She pocketed the drive as smoke grenades filled the corridor with dense gray clouds.

Phoenix Order operatives moved through the smoke with tactical precision. Ben dropped two with controlled bursts, but more kept coming. His movements were slowing, the virus and blood loss taking their toll.

"The window," Eva said, calculating distances. They were four stories up, but the laboratory's loading dock lay below, and its awning might break their fall.

"You go," Ben wheezed. "I'll hold them—"

"Together or not at all." She grabbed his vest, pulling him toward the shattered window as boots thundered closer.

That's when the screens in her mother's office blazed to life. Klaus's face appeared—not live, but a recording triggered by some hidden protocol.

"Eva," her digital father said, his expression softer than she'd seen in years. "If you're watching this, then you've discovered the truth about your mother. I want you to understand—I loved her until the end. But love and necessity are often at odds."

More operatives entered the smoke. Ben's weapon clicked empty.

The recording continued: "Your mother died believing she could save everyone. Such beautiful naivety. However, she never fully understood the scope of our work. The test sites in Lagos, Kinshasa, and Cairo aren't just proof of concept. They're already spreading. By the time you watch this, thousands will be dead. Millions within the week. The Phoenix rises from humanity's ashes."

"Monster," Eva spat at the screen.

"Perhaps," digital Klaus agreed, as if he'd anticipated her response. "But a necessary one. The genetic data we've collected shows humanity at a crossroads. Without intervention, our species faces extinction within three generations. The Phoenix Order offers controlled evolution. Yes, billions will die. But those who survive will inherit a world freed from genetic weakness, resource scarcity, and the burden of carrying evolutionary dead weight."

A figure emerged from the smoke—not a soldier, but Wilhelm von Hess himself, elegant even in tactical gear. "Touching family moment," he said, weapon trained on them. "But ultimately pointless. The virus has been released in twelve cities as of..." he checked his watch, "six minutes ago. The die is cast, if you'll forgive the classical reference."

Ben moved despite his wounds, stepping between Eva and the threat. Von Hess smiled and shot him twice in the chest.

"No!" Eva caught Ben as he fell, his weight driving them both to the floor. Blood spread beneath him, too much, too fast.

"How noble," von Hess observed. "The analyst finds redemption in sacrifice. Almost poetic. Your mother made similar gestures before Klaus had her poisoned. Tried to destroy her research, protect the world. Failed, of course. The weak always do."

Eva pressed her hands against Ben's wounds, feeling his life pumping out between her fingers. His eyes found hers, already glazing with shock. He tried to speak, blood bubbling from his lips.

"The micro-drive," von Hess continued conversationally. "Elena's final gambit. Yes, we know about it. Klaus found her research months ago and analyzed it thoroughly. Brilliant work, actually. A genetic mirror that would grant universal immunity. One small problem—it requires a living sample of the original viral strain. Not the weaponized versions we're releasing, but Elena's very first creation. And that sample was destroyed fifteen years ago."

The truth hit Eva like a physical blow. Her mother's final hope, the genetic mirror that could save everyone, was useless without the one component that no longer existed.

"Your father let you find those files," von Hess explained, enjoying her despair. "Let you hope. The Phoenix Order feeds on hope, you see. It makes the final surrender so much sweeter."

Ben's hand found Eva's, squeezing weakly. She looked down at him through tears, watching the light fade from his eyes. The man who'd protected her, loved her, believed in her when logic said to run, dying in her arms while humanity's extinction spread across the globe.

"Kill me," she whispered to von Hess. "I've lost everything. Just finish it."

"Oh no, my dear. You're far too valuable for that." He produced a set of restraints. "You see, your genetic profile is quite unique. Your mother's modifications during pregnancy, your father's enhancements—you're practically royalty of the Phoenix Order. You'll help us build the new world whether you want to or not."

More operatives entered the room, weapons trained on her. Ben's breathing had stopped, his hand going slack in hers. The virus samples were spreading globally. Her mother's research was useless. Her father's legacy of death was unstoppable.

Eva closed her eyes, cradling Ben's body, and for the first time in her life, she wanted to give up. The weight of failure crushed down on her, not just her own, but her mother's, Ben's, everyone who'd tried to stop this nightmare.

Von Hess stepped closer, confident in victory. "The old world dies today, Dr. Richter. And you'll have a front-row seat to watch the Phoenix rise from its ashes."

In her pocket, the micro-drive felt like a lead weight. Useless knowledge, pointless hope. Her mother's last gift had become her cruelest joke. Everything they'd fought for, everyone they'd lost—all for nothing.

The Phoenix Order had won.

And somewhere in her broken heart, Eva heard her mother's lullaby one last time. The melody that had promised safety, promised hope, promised that love could conquer the darkness.

All of it lies.

All of it ashes.

The world was ending, and Eva Richter was still alive to watch it burn.

# Chapter 27: The Dark Night

The maintenance tunnel stretched into darkness, a forgotten artery beneath the Alpine facility. Eva dragged Ben deeper into the shadows, her muscles screaming with each step. The blood trail they left would make them easy to track, but she had no choice but to follow it. The Phoenix Order forces would sweep the upper levels first, giving them perhaps an hour before the hunt moved underground.

She found an alcove where water pipes created a small sheltered space and carefully lowered Ben against the wall. His tactical vest was soaked with blood from von Hess's bullets. The armor had prevented immediate death, but the damage beneath was severe.

"Need to... keep moving," Ben gasped, his face pale in the emergency lighting's wan glow. "They'll find us..."

"Stop talking." Eva's hands shook as she cut away his vest. The entry wounds were high on the right side, missing his heart but catching lung tissue. Blood frothed with each labored breath. Without proper medical attention, he had at most a few hours.

The irony cut deep. She'd spent years developing viral delivery systems that could kill millions, but she couldn't save one man. The one man who mattered.

"Eva..." Ben caught her hand as she tried to fashion bandages from her jacket. "The drive. Your mother's research. You have to—"

"It's useless." The words came out bitter as poison. "Von Hess was right. We need the original viral sample, and it was destroyed years ago. My mother died for nothing. We're failing for nothing."

She pressed the makeshift bandages against his wounds, but blood seeped through immediately. The partial antidote they'd shared was losing its battle against the virus, and now Ben's body faced a war on two fronts.

"Not... nothing," Ben insisted. His hand found her face, thumb brushing away tears she hadn't realized were falling. "We destroyed their research. Set them back."

"They already released it! Lagos, Kinshasa, Cairo—thousands dead. Millions more coming." Eva's voice broke completely. "I created this. My work, my research, my naive belief that science could save the world. I handed them the tools for genocide."

The tunnel's silence pressed down on them, broken only by water dripping somewhere in the darkness and Ben's increasingly labored breathing. Above, she could hear boots on metal grating. The search had begun.

"Look at me," Ben whispered.

Eva met his eyes, seeing pain and determination mixed with something deeper.

"Syria," he continued with effort. "Forty-seven civilians. Wedding party. My analysis said the building held terrorists. I was wrong. Those faces... haunt me every night."

"That wasn't your fault. Bad intelligence—"

"I could have questioned it. Could have demanded better confirmation. But I was so sure, so confident in my abilities." He coughed, specks of blood hitting his lips. "The guilt nearly killed me. I'd have if I hadn't kept hiding from it. You saved me from that."

"I dragged you into worse," Eva protested. "You were safe in Berlin. Teaching, writing, away from all this violence. I brought death back into your life."

"You brought purpose." His grip on her hand tightened. "Eva, listening to you in that hotel bar, seeing your desperation to stop this nightmare—it reminded me why I joined the CIA in the first place. To protect people. To stand against evil."

"And look where it got you." She gestured helplessly at his wounds. "Dying in a tunnel while the world burns above us."

"Where it got me," Ben said slowly, each word a struggle, "is here with you. The woman who sacrificed everything to stop her father's madness. Who faced down the Phoenix Order when she could have run? Who chose to share the antidote rather than save herself?"

Eva shook her head. "None of it matters. They win. The virus spreads. The world reshapes itself into their sick vision of genetic purity. Everyone we tried to save—"

"Will know someone fought for them." Ben's eyes blazed with fierce conviction despite his weakening body. "That's what your mother understood. Why did she keep fighting even when Klaus had her cornered? Not because she could win, but because someone had to try."

The maintenance tunnel suddenly felt smaller, more intimate. Eva became hyperaware of every detail: the way Ben's blood mixed with condensation on the pipes, the distant hum of ventilation systems, the weight of the useless micro-drive in her pocket.

"I should have seen it," she whispered. "All those years working beside my father, and I never questioned deeply enough. Never pushed hard enough. If I'd discovered the truth sooner—"

"You'd be dead sooner." Ben's statement was matter-of-fact. "Klaus would have eliminated you like he did your mother. The Phoenix Order doesn't tolerate dissent."

"Maybe that would have been better. Without my research—"

"Someone else would have created it. The Phoenix Order has resources, brilliance, and determination. If not you, then another scientist. At least this way, you're here to fight back."

Eva laughed bitterly. "Fight back? I'm hiding in a maintenance tunnel with a dying man and useless data while twelve cities experience hell. Some fighter I am."

Ben's hand moved to her face again, forcing her to meet his gaze. What she saw there stole her breath. Not pity or resignation, but something that burned through her despair.

"Eva Richter," he said formally, as if making a declaration. "You are the bravest woman I've ever known. And I love you."

The words hung in the air between them, impossible and necessary.

"Ben—"

"No. Let me finish while I can." He struggled to sit straighter, grimacing at the movement. "I've spent three years hiding from the world, from myself, from any possibility of feeling again. Then you crashed into my life with your desperate mission, brilliant mind, and absolute refusal to give up. You made me remember who I used to be. Who I could be again?"

"A dead man," Eva said through tears. "That's what I made you."

"A man who mattered." Ben's voice grew stronger with conviction. "For the first time since Syria, my life has meaning. Purpose. Even if we fail here, even if the Phoenix Order reshapes the world, we tried. We stood against the darkness. That matters."

Above them, an explosion shook the facility. Dust rained from the tunnel ceiling as pipes groaned under the stress. The search was intensifying.

"They'll find us soon," Eva observed, surprised by her own calm.

"Then we make our time count." Ben pulled her closer, and she carefully arranged herself to avoid his wounds. "Tell me about the lullaby. Your mother's message."

Eva closed her eyes, letting the melody flow through her mind. "She used to sing it every night. Said it would keep the monsters away. I memorized every note, every word. When I found her research notes, I realized she'd encoded the genetic sequence in the lyrics. A, T, C, G—the building blocks of DNA hidden in a children's song."

"Clever woman."

"Too clever. The sequence is perfect, but without the original sample..." Eva trailed off, then suddenly sat up straight. "Oh my God."

"What?"

Her mind raced, pieces clicking together with terrible clarity. "My mother was paranoid about electronic surveillance, but she was also paranoid about physical security. She wouldn't have destroyed the only sample of the original virus. She'd have hidden it."

"Where?"

Eva thought back to her mother's office, her habits, and her quirks. The woman who hid micro-drives in coffee mugs, who encoded genetic sequences in lullabies, who built secret messages into everything she touched.

"The photo," she breathed. "On her desk. I'm building sandcastles. She always said it was her favorite, that it reminded her why she became a scientist—to build better worlds for children."

Ben stared at her. "You think she hid a viral sample in a picture frame?"

"Not in it. Behind it." Eva's excitement built despite their desperate situation. "She had a habit of pressing flowers, preserving them between glass plates. What if she preserved more than flowers?"

"That's insane."

"That's my mother." Eva felt hope kindle for the first time in hours. "Ben, she knew Klaus would eventually kill her. But she also knew he'd be too sentimental to clear out her office. She counted on his ego, his need to maintain the shrine to his victim."

Ben coughed again, more blood this time. "Even if you're right, we can't get back there. The facility's locked down, crawling with Phoenix Order forces."

"No," Eva said slowly, a plan forming. "But von Hess thinks he's won. He has me marked as valuable breeding stock for their new world. If I surrender, offer to cooperate..."

"Absolutely not." Ben's hand found hers with surprising strength. "Eva, no. They'll never let you near that office again."

"They will if I sell it right. If I break completely, convince von Hess that I've accepted defeat. He's arrogant enough to believe it, to want to show off his victory." She met Ben's eyes. "It's our only chance."

"Our chance? Eva, I'm dying. In an hour, maybe two, I'll be gone. This is your chance to run. Get to the authorities, warn them—"

"The authorities are compromised. The Phoenix Order has people every-where." Eva's jaw set with determination. "And you're not dying. Not on my watch."

"Eva—"

"No." She pressed her forehead against his, feeling his fever rising. "You told me you love me. Well, Ben Carter, I love you too. And I will not let you die in this tunnel. We're going to stop them. Together."

Ben's laugh turned into a cough. "How? I can barely sit up, and you're talking about infiltrating a facility full of people who want us dead."

"Because they don't want us dead. Not yet." Eva pulled back, her mind working through possibilities. "Von Hess wants me alive for breeding. You're leverage to ensure my cooperation. If I play this right…"

"You'll get yourself killed."

"Maybe. Or maybe I'll save the world." She carefully helped Ben to a more comfortable position. "My mother sacrificed herself, believing her work could make a difference. She was right—she just didn't live to see it through. But I can. We can."

The tunnel shook again as another explosion rocked the facility. Time was running out in more ways than one.

"Eva," Ben whispered, his strength fading. "Even if you find the sample, synthesize the counter-virus, deploy it globally—millions will still die. The virus is already spreading."

"Then we save who we can," Eva said fiercely. "We fight for every life, every chance, every hope. Because that's what makes us different from them. The Phoenix Order writes off billions as acceptable losses. We fight for every single soul."

Ben smiled despite his pain. "Your mother would be proud."

"She'd be terrified." Eva managed a watery smile in return. "But she'd do it anyway. Just like we will."

Footsteps echoed from the tunnel entrance. Multiple contacts, moving with military precision. Their time in the darkness was coming to an end.

Eva quickly kissed Ben, tasting blood and desperation and love. "Whatever happens next, remember—you gave me the strength to fight. You brought me back from despair. That matters more than you know."

"Eva—"

"I'm going to save you," she promised. "Save everyone I can. And when this is over, when the Phoenix burns and the world survives, we're going to find a quiet place and live boringly ever after. Deal?"

Ben's eyes shone with tears and admiration. "Deal."

The footsteps grew closer. Eva stood, raising her hands in surrender as tactical lights pierced the darkness. She had one chance to sell her broken spirit, to convince von Hess that she'd accepted defeat.

One chance to get back to her mother's office.

One chance to turn Elena Richter's final gift into humanity's salvation.

The Phoenix Order thought they'd won. They thought her broken, beaten, hopeless.

They were about to learn that Eva Richter had her mother's brilliance, her father's determination, and something neither of them had possessed—someone worth fighting for.

The dark night was ending.

It's time for the Phoenix to learn what real fire looks like.

# Chapter 28: The Cipher in the Song

The maintenance tunnel's darkness pressed against Eva like a living thing. Ben's breathing had grown shallow and irregular, each exhale weaker than the last. She'd torn strips from her shirt to supplement the makeshift bandages, but blood still seeped through. The virus and his wounds were winning their dual war against his body.

"Stay with me," she whispered, cradling his head in her lap. The emergency lighting cast everything in hellish red, making the blood on her hands look black. "Just a little longer."

Ben's eyes fluttered open, focusing with effort. "Eva... the lullaby. Sing it."

"What?" She thought the fever had taken his mind. "Ben, this isn't—"

"Please." His hand found hers, grip barely there. "Want to... hear something beautiful? Before..."

Before the end. The words he couldn't say hung between them like a blade.

Eva's throat constricted, but she began to sing, her voice echoing softly in the tunnel:

*"Kleine Sterne leuchten hell,Durch die Nacht wie Muschelschell,Adenin tanzt mit Thymin,Cytosin mit Guanin..."*

The German words came naturally, thanks to muscle memory from thousands of bedtime stories. Little stars shine brightly through the night like seashells. Adenine dances with thymine, cytosine with guanine...

She'd always thought the last lines were her mother's poetic way of making science magical for a child. The base pairs of DNA transformed into dancing partners in the night. But as she sang, something nagged at her consciousness.

*"Wenn der Phosphor bindet fest, Baut die Helix sich ihr Nest, Drei-fünf-drei und fünf-drei-Ende, Drehen sich wie Liebeshände..."*

When the phosphorus binds tight, the helix builds its nest, three-five-three and five-three ending, turning like lovers' hands...

Eva stopped mid-verse, her mind racing. The song wasn't just incorporating scientific terms for whimsy. The entire structure was precise and methodical. Phosphorus bonds. Helix formation. Directional notation of DNA strands.

"My God," she breathed.

Ben's eyes had closed again, but a slight smile touched his lips. "Figured it out?"

"She didn't encode the kill switch sequence in the lullaby." Eva's voice shook with revelation. "She encoded instructions for reconstructing it. The song isn't the message—it's the blueprint."

Her mother had been even more brilliant than she'd realized. Rather than risk a static genetic sequence that might be discovered or corrupted, Elena Richter had hidden a methodology. A way to rebuild the counter-virus from first principles, using the lullaby as a guide.

Eva gently laid Ben's head on her folded jacket and pulled out the micro-drive from her mother's office. In her panic, she'd assumed it contained the kill switch itself. But as she held it up to the emergency lighting, she noticed something etched into its surface. Musical notes. The first bar of the lullaby.

"She knew," Eva whispered. "She knew Klaus would find any hidden research, any static data. So, she made it dynamic. The lullaby provides the structure, but the key..."

She closed her eyes, forcing herself to think like her mother. Elena had been both paranoid and sentimental. Everything had layers of meaning, especially things connected to Eva.

The answer came like lightning.

"My genome," she said aloud. "The modifications she made during pregnancy. She didn't just enhance my immune system or cognitive function. She embedded something in my DNA itself."

Ben coughed, blood speckling on his lips. "Eva..."

"No, listen!" She grabbed his hand, energy coursing through her despite exhaustion. "The lullaby tells me how to read it. Three-five-three and five-three ending—that's the direction to read DNA strands. The dancing base pairs aren't random; they're specific sequences to look for. And the chorus..."

She sang again, this time hearing the hidden message:

*"Schlaf mein Kind, die Welt wird neu, In dir liegt der Hoffnung Treu, Wenn die Phönix steigt empor, Öffnet sich das Lebenstor..."*

Sleep, my child, the world renews, in you lies hope so true, when the Phoenix rises high, opens up life's gateway...

"In your lives, hope," Eva repeated. "She literally meant IN me. The counter-virus sequence is encoded in my own DNA, as a result of the modifications she made. The lullaby is the cipher to extract it."

But knowing this created a new problem. To decode her own genome and reconstruct the counter-virus, she needed sophisticated equipment. The kind found only in the main laboratory, now crawling with Phoenix Order forces who thought her captured or dead.

"Have to go back," she said, already knowing Ben would protest.

"Eva, no." His voice was barely a whisper. "Too dangerous. Find another way."

"There is no other way." She checked his wounds again. The bleeding had slowed, but his skin was burning with fever. "The synthesis equipment, the genetic sequencers, the viral samples—everything we need is up there."

"They'll kill you."

"Maybe." Eva pulled the micro-drive's chain over her head, wearing it like a talisman. "But von Hess wants me alive for their breeding program. That gives me an edge. If I can convince them I'm broken, defeated..."

Ben tried to sit up, but failed. "I can't protect you."

"You already have." She kissed his forehead, tasting salt and copper. "You kept me from giving up. Reminded me that the fight matters, even if we lose. That's all the protection I need."

From somewhere above came the sound of machinery. The Phoenix Order was relocating equipment, likely in preparation for a global deployment. Time was running shorter by the minute.

Eva quickly reviewed what she knew. The lullaby was a cipher. Her DNA contained the encoded counter-virus. The microdrive likely held additional instructions or processing algorithms. Together, they formed her mother's final gift—not just a cure, but a way to ensure it could never be fully suppressed.

"Brilliant," she murmured, appreciating the elegant paranoia of it all. Even if the Phoenix Order captured the data, they couldn't use it without Eva herself. And even if they had Eva, they wouldn't know how to look in her own genetic code without the lullaby's guidance.

"What about the original virus?" Ben asked, fighting to stay conscious. "Von Hess said... the sample was destroyed."

Eva considered this. Her mother's plan seemed complete except for that crucial element. Then she remembered the photo on the desk. Her five-year-old self building sandcastles, preserved behind glass like pressed flowers.

"The beach," she said suddenly. "That photo was taken at the Baltic Sea, two months after my mother started her research. She always said that day was special, that it represented hope for the future."

"You think she hid a sample at a beach?"

"No." Eva's mind raced through possibilities. "But what if the sample isn't hidden in a place? What if it's hidden in a person?"

Ben's eyes widened with understanding. "You. She used you as a living repository."

"The modifications during pregnancy. Everyone assumed they were enhancements, but what if some were preservation? Dormant viral sequences integrated into my genome, inactive but intact." Eva felt pieces clicking together. "That's why the Phoenix Order marked me as breeding stock. They don't know about the counter-virus, but they detected something unique in my genetics."

It was elegant and terrifying. Elena Richter had transformed her own daughter into both the lock and the key for humanity's salvation. Everything needed to stop the Phoenix Order existed within Eva herself.

"Help me understand," Ben said, his analytical mind still working despite his body's deterioration. "You have the counter-virus encoded in your DNA. The lullaby tells you how to extract it. The original virus is dormant in your system. But how do you synthesize all this into something usable?"

Eva pulled out her mother's microdrive again, examining it more carefully. Beyond the musical notation, she noticed microscopic text etched around the edge. Chemical formulas. Processing sequences. Temperature gradients.

"The synthesis protocol," she breathed. "It's all here. The micro-drive contains the technical instructions. The lullaby provides the genetic cipher. My DNA holds both virus and cure. I just need the equipment to bring it all together."

"And an army between you and that equipment."

"Not an army. Just men who think they've already won." Eva stood, decision crystallizing. "Von Hess's arrogance is their weakness. He'll want to gloat, to show off their victory. I'll give him that chance."

She looked down at Ben, memorizing his face. Even bloodied and pale, he was beautiful to her. The man who'd chosen hope over safety, who'd given her strength when despair seemed the only option.

"I'll come back," she promised. "With the cure and medical supplies. You just have to hold on."

"Eva..." He caught her hand as she turned to go. "Your mother would be proud. You've become everything she hoped."

"Not yet." Eva squeezed his fingers gently. "But I will be."

She moved toward the tunnel entrance, where the voices of the Phoenix Order grew louder. Her mother's gift burned against her chest—not just the micro-drive, but the knowledge that Elena Richter had thought of everything. The paranoid brilliance that had hidden a cure in a lullaby, a virus in a daughter, and hope in the darkest corners of human ambition.

The voices were closer now. Eva could make out von Hess commanding the search teams. In moments, they would find this tunnel, Ben. She had to act first.

"I surrender!" she called out, stepping into the harsh glare of tactical lights. "I'll cooperate. Just... please. He's dying. I'll do whatever you want."

Von Hess himself appeared from behind the security team, his satisfaction evident. "Dr. Richter. How practical of you to finally see reason."

"He needs medical attention," Eva said, letting real desperation color her voice. "Please. Save him, and I'll help you with whatever you need. My research, my knowledge, myself—it's all yours."

"Touching." Von Hess gestured to his men. "Bring the analyst. Keep him alive for now—he may prove useful leverage."

As soldiers moved past her into the tunnel, Eva forced herself not to react. This was the gambit. Play the broken woman, the defeated scientist. Let them think grief and fear had shattered her will.

But inside, her mother's lullaby played on repeat, each verse revealing new secrets. The gift Elena Richter had hidden wasn't just scientific knowledge—it was faith that her daughter would be strong enough to use it.

The Phoenix Order had made one crucial mistake in their grand plan for human evolution.

They'd forgotten that mothers protecting their children were the most dangerous force in nature.

And Elena Richter had protected hers across decades, beyond death itself, with a song that would save the world.

# Chapter 30: The Final Approach

T he Phoenix Order's ceremonial hall occupied what had once been the facility's main conference center, now transformed into a space that was part cathedral and part war room. Black banners bearing the golden phoenix emblem hung from the vaulted ceiling. At the same time, holographic displays showed countdown timers for each global release point. Two hundred and seventeen members had gathered, their faces reflecting the fervor of true believers witnessing the birth of their new world.

Eva stood among them, wearing the black ceremonial robe someone had thrust into her hands. The fabric felt heavy with significance, or perhaps that was just the weight of the concealed micro-drive against her chest. Around her, geneticists who'd spent years preparing for this moment whispered excitedly. She recognized many faces—former colleagues who'd been Phoenix Order sleepers, their true allegiance hidden behind academic credentials and false ethics.

"Brothers and sisters," Wilhelm von Hess's voice resonated through the hall as he took the podium. Behind him, a massive holographic globe rotated slowly, each release point pulsing like a heartbeat. "Today, we correct nature's course. Today, we save humanity from itself."

Applause rippled through the gathering. Eva joined in, playing her part while her eyes tracked the security positions. Six guards are stationed at the main entrance, four at the emergency exits, and two flank the podium. And somewhere in the medical bay, Ben lay under minimal guard—von Hess's arrogance assuming a dying man posed no threat.

"For too long," von Hess continued, his charisma filling the space, "we have watched our species weaken. Medical advances are preserving defective genes. Social policies encourage the multiplication of the unfit. Political correctness is preventing us from speaking nature's truth—that all humans are not created equal."

Murmurs of agreement rose from the assembly. Eva felt bile rise in her throat, but forced herself to nod along. Eight minutes until the synthesis is completed. Seven until global deployment. The margins were impossibly thin.

"But we," von Hess gestured to encompass them all, "we had the courage to act. Under Klaus Richter's brilliant leadership, we developed the tools of selective evolution. His daughter's work—" his eyes found Eva in the crowd, "—provided the final piece. Even in her misguided rebellion, she served our cause."

Eva lowered her head as if in shame, but she was actually checking the small communicator Ben had slipped her during their brief moment in the deployment center. His message was simple: "Ready. Signal when."

"Some call us monsters," von Hess's voice rose with passion. "They're right. We are monsters to the parasites. We are devils to the weak. We are death itself, to the genetic refuse cluttering our planet. But to the future? To the strong who will inherit our new world? We are saviors. We are gods!"

The assembly erupted in cheers. Scientists who'd dedicated their lives to understanding genetics had twisted that knowledge into justification for mass murder. Eva wondered how many had started like her, believing they were helping humanity, only to be seduced by the promise of power over evolution itself.

A new display materialized above von Hess—live feeds from the release points. Eva saw operatives in position at JFK Airport, Heathrow, Tokyo Station, and dozens of other locations. Each carried innocuous-looking devices—perfume bottles, air fresheners, small mechanical items that would release invisible death into ventilation systems.

"Observer reports ready at all primary sites," a technician announced. "Secondary sites confirm standby status. Tertiary backups armed and waiting."

Von Hess smiled like a proud father. "Then let us begin with the traditional reading."

An elderly woman, identified as Dr. Margaret Steinberg, a Nobel laureate in genetics, approached the podium, carrying an ornate book. Her voice, once used to lecture on the wonders of DNA, now recited the Phoenix Order's twisted scripture:

"From the ashes of the old, the new shall rise. From the death of the many, the few shall thrive. Nature demands sacrifice. Evolution requires selection. We are the instruments of both."

"We are the instruments," the assembly repeated in unison.

Eva mouthed the words, her mind racing. Six minutes to synthesis completion. Five to deployment. She needed to move, but too early would expose her. Too late would doom millions.

"Dr. Richter," von Hess called out suddenly. "Eva. Come forward."

Her blood froze, but she had no choice. The assembly parted as she walked to the podium, feeling two hundred pairs of eyes tracking her movement. Some held sympathy for Klaus's daughter. Others suspected the traitor among them.

"Ladies and gentlemen," von Hess announced, placing a possessive hand on Eva's shoulder, "behold the paradox of evolution. A brilliant mind trapped by outdated morality. But genetics will out. Dr. Richter has returned to us, accepting her role in our new world."

He handed her a ceremonial device—a golden trigger that would signal the simultaneous global release. "Your father dreamed of this moment. He would want you to have this honor."

Eva took the device with trembling hands, not from fear, but from rage carefully disguised as emotion. They wanted her to begin the genocide. To put her finger on the trigger that would kill billions.

"I..." she began, then stopped as if overcome. "I need a moment. To honor my father properly."

Von Hess's eyes narrowed slightly, but he nodded. "Of course. We have four minutes. Compose yourself."

Eva stepped back from the podium, mind racing. Four minutes. The synthesis would be completed in five. Even if she stalled, deployment would begin before the counter-virus was ready. Unless...

She activated Ben's communicator with a subtle touch. His response was immediate—an explosion rocked the facility's east wing.

"Security breach in medical!" someone shouted. "The analyst—he's escaped!"

Von Hess's face contorted with fury. "Impossible. He was dying."

Another explosion, this one closer. The ceremonial hall's lights flickered as backup power engaged. Several assembly members moved toward the exits, but von Hess's voice cracked like a whip:

"No one leaves! This is a minor distraction. Security will handle one wounded man." He turned to Eva. "Your lover's heroics change nothing. The deployment proceeds on schedule."

But Eva saw the doubt creeping into some faces. Ben's "escape" had shattered the moment's solemnity, reminding them that opposition still existed. She pressed the advantage:

"He's heading for the synthesis lab," she said, loud enough for all to hear. "He knows about the counter-virus work. If he destroys it—"

"Then we proceed without optimization," von Hess snapped. "The primary deployment is all that matters. Dr. Richter, return to the podium. Now."

She obeyed, buying precious seconds with slow steps. Three minutes to deployment. Four to synthesis completion. Ben's diversions had bought time, but not enough.

That's when she noticed something on the holographic displays. Several operatives at release sites were touching their ears, receiving communications. Their confident postures shifted to uncertainty.

"Sir," a technician called out, voice tight with concern. "We're receiving reports of security alerts at multiple locations. Local authorities are responding to anonymous tips about suspicious devices."

Von Hess whirled on the technician. "How many sites?"

"Seventeen and climbing. Someone's leaked our positions."

Eva hid her surge of hope. Ben hadn't just escaped—he'd managed to transmit warnings. Not enough to stop everything, but enough to disrupt the perfect synchronization the Phoenix Order required.

"Accelerate the timeline," von Hess commanded. "Deploy immediately at all uncompromised sites."

"Sir, without synchronization, the infection patterns—"

"Will be less elegant but equally effective." Von Hess turned back to Eva, his composure cracking. "Trigger the release. Now."

Eva held the golden device, feeling the weight of the world literally in her hands. Around her, the assembly waited. On the screens, operatives prepared to release death into the air that billions breathed.

Two minutes to synthesis completion.

One hundred and ninety-four sites are still operational.

Millions of lives hang in the balance with her next decision.

"For my father," she said quietly, raising the device. "For Klaus Richter, who dreamed of a better world."

Her finger hovered over the trigger. The assembly leaned forward in anticipation. Von Hess smiled in triumph.

Then Eva smiled too. "He was wrong."

She hurled the device at the main holographic projector, the golden trigger smashing into delicate electronics. Sparks erupted as the displays flickered and died. In the confusion, she dove for the emergency alarm, slamming her palm against it.

Sirens wailed throughout the facility. "Contamination breach. All personnel must evacuate immediately. This is not a drill."

The assembly erupted in chaos. Some ran for the exits. Others stood frozen in disbelief. Security moved toward Eva, but the evacuation protocol had automatically unlocked all doors, and panicked Phoenix Order members stampeded through.

Von Hess's voice rose above the chaos: "Stop her! The deployment must proceed!"

But Eva was already moving, using the crowd as cover. She had to reach the synthesis lab. The counter-virus would be complete any second, but without her to input the distribution commands, it would sit useless while the world died.

A hand grabbed her robe. She spun to find Dr. Steinberg, the elderly geneticist, eyes blazing with fanatic fury. "You've doomed us all! Natural selection will proceed without guidance! Chaos instead of control!"

Eva wrenched free. "Better chaos than your genocide!"

She burst from the ceremonial hall into corridors flooded with evacuating personnel. The contamination alarm had triggered full emergency protocols. Even loyal Phoenix Order members wouldn't risk exposure to their own weapons.

Her watch showed ninety seconds to completion of synthesis. She ran faster than she'd ever run, her mother's micro-drive bouncing against her chest with each stride. Behind her, she heard von Hess shouting orders, trying to restore control.

The synthesis lab doors came into view. Almost there. Almost—

Gunfire erupted. Bullets sparked off the walls as Eva dove through the doorway. Von Hess himself pursued, weapon drawn, face twisted with rage.

"You think you've won?" he snarled, advancing as Eva scrambled toward the synthesis chamber. "Even if you've created a counter-virus, how will you distribute it? We control the networks, the infrastructure. You're just one woman with a test tube!"

The synthesis chamber chimed—process complete. A single vial of clear liquid sat in the collection port. Elena Richter's legacy made real.

Eva grabbed the vial as von Hess fired again. She rolled behind equipment as bullets shattered glass and sparked off metal. Sixty seconds since deployment began at the uncompromised sites. Already, invisible death was spreading through ventilation systems, carried on the breath of unknowing victims.

"It's over!" von Hess shouted. "Submit to genetic destiny!"

Eva's hand found something—a pressurized dispersal unit, designed for aerosol testing. Her mother's voice echoed in her memory: "Sometimes the simplest solutions are the most elegant."

She loaded the counter-virus into the dispersal unit and adjusted the settings for maximum coverage. It wouldn't reach everyone, but in a confined space...

"The facility's ventilation system," she called out to von Hess. "It's still in emergency mode, circulating air at maximum capacity to clear contamination."

Understanding dawned in his eyes. "You wouldn't. The concentration would need to be—"

"Approximately what I'm about to release." Eva stood, holding the dispersal unit. "Your people will be the first exposed to the counter-virus. They'll carry it out of here and spread it to others. Not perfect distribution, but a start."

"Our virus is already deployed!"

"Yes," Eva agreed. "But viruses mutate. Spread. Interact. My mother designed the counter-virus to be more aggressive than yours. It will find your weapon wherever it goes and neutralize it. Maybe not in time to save everyone, but enough will survive. Humanity will endure."

Von Hess aimed at her heart. "Then you'll die knowing you failed."

"No," said a voice from the doorway. "She won't."

Ben stood there, bloodied but unbowed, holding a security officer's weapon. His shots were precise—von Hess's gun flew from his hand as bullets caught his shoulder and leg, dropping him to the floor.

"Ben!" Eva moved toward him, then remembered the dispersal unit in her hands. The counter-virus. She had to release it now, while there was still time.

"Do it," Ben said, understanding. "Save them all."

Eva activated the dispersal unit. A fine mist erupted, invisible but for the way it caught the emergency lighting. The counter-virus, finally free, began its work. Within seconds, it would saturate the facility. Within hours, it would spread beyond.

Von Hess laughed through his pain. "Too little, too late. The Phoenix rises regardless."

"Maybe," Eva said, helping Ben to a chair as sirens continued to wail. "But my mother taught me something your philosophy never understood. Life finds a way. Not the strongest life, not the purest, but life itself. Diverse, chaotic, beautiful life."

On the screens that still functioned, she could see the first reports coming in. Mysterious illnesses at several airports. Emergency responses in multiple cities. The Phoenix Order's virus had been released.

But so had Elena Richter's final gift. The race was on—death versus cure, selection versus salvation.

Eva held Ben's hand as they watched the world begin to change. They'd done everything they could. Now humanity would decide its own fate, as it always had.

Not through engineered selection, but through the messy, miraculous process of survival.

The Phoenix had risen.

But from its ashes, hope took wing.

# Chapter 29: The Phoenix Rises

The Phoenix Order's medical bay was a stark contrast to the squalor of the maintenance tunnel. White walls, surgical steel, the antiseptic smell of a place where lives were saved—or transformed into something monstrous. Eva watched through the observation window as medics worked on Ben, their efficiency born from practice on test subjects.

"Remarkable constitution," von Hess observed beside her. "Most men would be dead from those wounds. Your analyst has surprised us all."

Eva kept her expression carefully neutral, playing the part of the defeated woman. They'd cleaned his wounds, given him blood, and stabilized his vitals. Not from mercy, she knew, but because von Hess understood the value of leverage. As long as Ben lived, Eva would cooperate.

"The virus in his system complicates treatment," the lead medic reported through the intercom. "We've administered suppressants, but without the specific antidote..."

"He has perhaps six hours," von Hess finished. "Sufficient time for Dr. Richter to demonstrate her newfound cooperation." He turned to Eva with a smile that never reached his eyes. "Shall we discuss your mother's research?"

Eva nodded meekly, but her mind was racing. Six hours. The global deployment would begin in four. She had to move faster than anticipated.

They walked through corridors she knew by heart, past laboratories where she'd spent her youth believing she was saving the world. Phoenix Order personnel nodded respectfully to von Hess, but their eyes on Eva held contempt. Klaus Richter's traitorous daughter was finally brought to heel.

"Your mother was brilliant," von Hess said conversationally. "Pity she lacked vision. Elena could have been one of our greatest assets. Instead, she forced us to make her a martyr."

"She was naive," Eva agreed, hating herself for the words even as she sold the performance. "She never understood that progress requires sacrifice."

"Precisely! Your father grasped this truth. Evolution is not kind, Dr. Richter. It is efficient." Von Hess gestured to a wall display showing global infection projections. "In seventy-two hours, the weak will begin dying. In a week, only the genetically superior will remain. Humanity's next chapter begins with necessary cruelty."

They reached the main laboratory, which had now been restored after the chaos of the test subject's release. Eva saw that her father's blood had been cleaned from the floor, and his body had been removed. Only the work remained—terminals glowing with viral sequences, deployment schedules, the machinery of genocide humming along smoothly.

"I need access to the synthesis equipment," Eva said, keeping her voice steady. "And my mother's office. If I'm going to help optimize the deployment, I need to understand her original work."

Von Hess studied her carefully. "Your sudden cooperation is... interesting. What guarantee do I have that you won't attempt sabotage?"

Eva met his gaze, letting him see carefully crafted desperation. "Ben is dying. My father is dead. Everything I believed in has been a lie. What's left for me except survival?" She paused, then added the words she knew he wanted to hear. "The world you're creating will need scientists. I want to be part of it."

"Ambition replacing idealism. How refreshing." Von Hess nodded to the guards. "Grant her supervised access. If she attempts anything foolish, kill the analyst immediately."

Eva bowed her head in acquiescence, hiding the fury in her eyes. Let him believe he'd broken her. Let them all believe it. Her mother had hidden resistance in submission for years before Klaus discovered her betrayal.

The synthesis laboratory hadn't changed much since Eva's last legitimate work here. Banks of genetic sequencers, viral cultivation chambers, molecular assembly units—everything needed to create or destroy a pathogen. Under the watchful eyes of two guards, she began her performance.

"I need to analyze my own genetic markers," she explained, drawing her blood with practiced ease. "My mother's modifications made me immune to certain viral strains. Understanding how might help optimize the deployment."

It wasn't entirely a lie. The guards, chosen for their loyalty rather than scientific knowledge, didn't question her as she fed her blood into the sequencer. The machine hummed to life, beginning its analysis of her DNA.

While it worked, Eva moved to her mother's office. The door stood open now, Klaus's sentimental preservation ended with his death. But the photo remained on the desk—five-year-old Eva building sandcastles, her mother's shadow falling across the beach.

"Childhood memories?" von Hess asked from the doorway. He'd followed her, suspicious despite her apparent cooperation.

"Trying to understand her thought processes," Eva replied, picking up the frame. "She always said this day was special. I thought it was sentiment, but now I wonder if there was more."

She opened the frame carefully, as if examining it for the first time. Behind the photo, pressed between glass plates like dried flowers, were what appeared to be small crystalline structures. To others, they might appear like sand or salt crystals. But Eva recognized the preservation medium her mother had developed—a way to maintain biological samples indefinitely.

"Sand from the beach?" von Hess mused. "Your mother's sentimentality seems intact."

"Yes," Eva agreed, pocketing the crystals carefully. "Just sand."

But she could feel the weight of them, the dormant viral samples her mother had hidden in plain sight for fifteen years. The original strain, preserved at the moment of creation, before Klaus's modifications transformed it into a weapon.

They returned to the synthesis lab where the genetic sequencer had completed its initial analysis. Eva pulled up the results, her breath catching at what she saw. Her mother hadn't just hidden instructions in her DNA—she'd created an entire biological library.

"Fascinating," she murmured, genuinely amazed as she traced the sequences. "The modifications go deeper than we thought. Look here—dormant sequences that could be activated under specific conditions."

Von Hess leaned closer, intrigued despite himself. "Your mother turned you into a living repository?"

"More than that." Eva's fingers flew across the keyboard, following the lullaby's cipher. "She created a biological lock. These sequences can only be activated in a specific order, using specific triggers. It's like..." She paused, pretending to realize something. "It's like she was planning for someone to need this information later."

"The counter-virus," von Hess said sharply. "She hid it in you?"

Eva let fear creep into her voice. "I don't know. Maybe. But without understanding her encryption method..."

"Decrypt it." His voice turned cold. "You have two hours before the deployment begins. If you haven't produced results by then, we'll test exactly how much pain the analyst can endure before death."

Eva nodded quickly, turning back to her work. Two hours. She could work with that.

As von Hess left to oversee deployment preparations, Eva began the real work. The lullaby played in her mind as she isolated specific genetic sequences. Adenine dancing with thymine, cytosine with guanine. Each verse revealed another piece of the puzzle.

She carefully dissolved the preserved viral samples in solution, watching the original strain come alive under the microscope. So simple compared to its weaponized descendants. So elegant in its basic structure. Her mother had started with something that could have been beneficial—a virus designed to enhance immune responses. Only Klaus's ambition had transformed it into an instrument of genocide.

The guards watched her work with bored incomprehension. To them, she was just another scientist following orders. They didn't notice her preparing two separate synthesis chains—one that appeared to optimize the Phoenix Order's virus, another hidden in subroutines that followed her mother's design.

An hour passed. Eva's hands moved with practiced precision, but her mind was with Ben. Was he still alive? Still fighting? She had to believe he was, had to believe there would be someone left to save when this was over.

The genetic compiler beeped, indicating it had successfully mapped her modified DNA. Eva pulled up the results, following the lullaby's guidance to extract the hidden sequences. There—buried in what appeared to be junk code—was her mother's true gift.

Not just a counter-virus, but something far more elegant. A genetic mirror that would turn the Phoenix Order's weapon against itself. Every variant they'd created, every targeted sequence designed to kill specific populations—the mirror would reverse them all. Instead of identifying victims, the virus would identify itself and trigger self-destruction.

"Brilliant, mother," Eva whispered. "Absolutely brilliant."

She began the synthesis, carefully combining the original viral sample with the extracted sequences. The process would take forty minutes, cutting it close to von Hess's deadline. But she needed every second to ensure accuracy. One mistake and she'd create another weapon instead of a cure.

As the machines worked, Eva prepared her backup plan. She uploaded the synthesis protocol to a hidden partition in the laboratory's network, encrypted with her mother's lullaby. If she failed here, if the Phoenix Order stopped her, at least the information would exist somewhere.

"Dr. Richter." One of the guards had received a communication. "Mr. von Hess requires your presence in the deployment center."

"I can't leave now," Eva protested. "The synthesis is at a critical stage—"

"He insists." The guard's hand moved to his weapon. "Immediately."

Eva glanced at the synthesis progress—twenty-three minutes remaining. She had no choice but to comply, praying the automated systems would complete the process without her.

The deployment center sprawled across the facility's top floor, a high-tech command hub worthy of a military operation. Which, Eva realized, was exactly what this was. Screens showed Phoenix Order personnel at airports, train stations, and water treatment facilities worldwide. Each operative carried sealed containers, waiting for the synchronized release command to be given.

"Ah, Dr. Richter." Von Hess stood at the central console like a conductor preparing his orchestra. "I wanted you to witness history. Your work made this possible. You should see it implemented."

Ben was there, strapped to a medical gurney, barely conscious but alive. Eva's heart leaped even as she maintained her defeated facade.

"Fifteen minutes to deployment," a technician announced. "All stations report ready."

"You see?" Von Hess gestured to the screens. "Perfect coordination. While governments debate and populations remain ignorant, we act. By the time anyone understands what's happening, the new world order will be inevitable."

Eva watched the countdown timer, calculating frantically. The synthesis would complete in approximately eighteen minutes. Three minutes after deployment began. Too late to stop the initial release, but perhaps in time to minimize casualties if she could transmit the cure globally.

"It's magnificent," she said, playing her role while her mind raced through possibilities.

"Your father thought so too." Von Hess pulled up Klaus's final video logs. "He was quite proud of you, you know. Even at the end. Shall we watch together?"

On screen, Klaus appeared in the same laboratory where he'd died. However, this recording was made recently, just hours before their confrontation.

"My dear Eva," her father's image said. "If you're watching this, then our reunion has likely ended badly. I want you to know I forgive you. Your betrayal was inevitable—you have too much of your mother in you. But genetics will out, my daughter. The modifications Elena made ensure you'll survive what's coming. You'll live to see the world we're creating, whether you wish to or not."

Eva felt sick. Even in death, Klaus believed his vision would triumph.

"He loved you," von Hess observed. "In his way. He even argued for leniency after your escape. I overruled him, of course. Sentiment is weakness."

"Yes," Eva agreed through gritted teeth. "Weakness."

Ten minutes to deployment. The synthesis would be complete in thirteen. She needed a distraction, a way to buy time.

That's when she noticed Ben's eyes were open, alert despite his apparent unconsciousness. He'd been watching, listening, waiting. Their gazes met for a fraction of a second—enough for understanding to pass between them.

"Mr. von Hess," Eva said suddenly. "There's something you should know about my mother's work. Something my father never discovered."

The countdown reached nine minutes. Around the world, Phoenix Order operatives prepared to release humanity's doom.

And in a small synthesis chamber three floors below, Elena Richter's final gift neared completion.

The Phoenix was rising.

But from its ashes, something unexpected would emerge.

Hope.

# Chapter 31: Synthesis

The Alpine facility's emergency sirens created a cacophony that masked the sound of gunfire echoing through its corridors. Eva clutched the dispersal unit's empty canister, her mind racing through distribution calculations. The counter-virus was now in the air system, but that was only the beginning. Local saturation wouldn't stop a global pandemic. She needed something more—a way to replicate and spread the cure faster than the Phoenix Order's weapon.

"Can you walk?" she asked Ben, who leaned heavily against the synthesis chamber.

"I'll manage." His face was pale, but his eyes held the same determination that had carried them this far. "What's the plan?"

"The facility's bio-defense network." Eva moved to the main terminal, her fingers flying across the interface. "It's connected to WHO databases, CDC systems, and international health monitoring networks. If I can upload the counter-virus synthesis data..."

"They'll think it's an attack," Ben warned, checking his appropriated weapon. "Phoenix Order will have people in those organizations."

"Not everywhere. Not everyone." Eva initiated the upload sequence, embedding her mother's lullaby as a verification key. "Some will recognize the code for what it is—salvation."

Von Hess groaned from the floor, his blood pooling beneath him. "Naive girl. You think posting formulas online will save anyone? The infrastructure is already compromised. Communications are being cut. Governments are initiating lockdowns."

"Shut up," Ben growled, but Eva heard the truth in von Hess's words. The Phoenix Order had planned for interference. They'd have protocols to limit information spread, to control the narrative.

"He's right about one thing," Eva said, pulling up global network diagnostics. "Look. Major communication hubs are going dark. They're isolating infection zones to prevent panic and information flow."

On the screens, red zones expanded across the world map like spreading blood. New York, London, Tokyo, Berlin—all showing signs of the virus taking hold. The death toll would start small and escalate exponentially. Within hours, hospitals would be overwhelmed. Within days, the infrastructure would collapse.

"There has to be another way," Ben said, moving to cover the laboratory entrance despite his injuries.

Eva's mind raced through possibilities. Then she remembered something from her childhood, watching her mother work. Elena Richter hadn't just been paranoid about digital surveillance—she'd been creative about circumventing it.

"The maintenance network," Eva breathed. "Every major facility built in the last thirty years has automated maintenance systems. They communicate through separate channels, updating each other about part requirements, system failures, and optimal protocols."

"So?"

"So they're invisible. Nobody monitors maintenance networks for security threats because they only carry diagnostic data." Eva accessed the facility's maintenance hub, diving deep into its communication protocols. "But data is data. If I can embed the synthesis information in maintenance packets..."

Ben's weapon snapped up as footsteps approached. "Company."

The first Phoenix Order security team rounded the corner into Ben's firing solution. His shots were economical and precise despite his wounds. Two men dropped before the others pulled back, laying down suppressing fire.

"How long do you need?" Ben asked, ejecting a spent magazine.

"Ten minutes. Maybe fifteen." Eva's hands never slowed on the keyboard. "The maintenance network connects to thousands of facilities globally. Hospitals, research centers, universities—places with the equipment to synthesize the cure."

"I'll buy you twenty." Ben checked his ammunition. "But Eva... there are too many of them. Eventually—"

"Don't." She looked at him, memorizing his face. "We're not dying here. Not after everything."

He smiled, the expression transforming his pain-etched features. "Yes, ma'am."

The security team tried another approach, lobbing flash-bang grenades into the laboratory. Ben kicked one back, using the explosion's disorientation to advance on their position. Eva heard the firefight intensify but forced herself to focus on the code.

The maintenance network was beautifully simple. Every connected facility received regular updates about filter specifications, chemical requirements, optimal temperature ranges. Eva began embedding the counter-virus data into these mundane packets. To automated systems, it would look like routine information. But to any human who examined the data...

"Warning," she typed in multiple languages. "Critical health information. Cure for Phoenix Virus. Synthesis instructions follow."

She included everything—her mother's research, the lullaby cipher, the synthesis protocols. Even instructions for extracting the necessary components from standard laboratory supplies. The elegant complexity of Elena Richter's work is reduced to its essential elements, making it simple enough for any competent lab to reproduce.

"Eva!" Ben's shout made her look up. He'd been driven back into the laboratory, blood seeping through fresh wounds. Behind him, Phoenix Order forces advanced with military precision.

But something was wrong with their formation. Some operatives moved sluggishly, their tactical coordination failing. Others had removed their helmets, gasping for air.

"The counter-virus," Eva realized. "It's affecting them."

The Phoenix Order had engineered their weapon to ignore certain genetic markers—those of their own. But Elena's counter-virus recognized no such distinctions. It targeted the Phoenix weapon itself, turning the virus against its creators. Those who'd been exposed to early variants during testing were feeling the effects first.

"Poetic justice," Ben gasped, taking position behind an overturned equipment rack.

Von Hess laughed weakly from the floor. "You think this changes anything? The strong will adapt. The weak will die. Evolution continues regardless."

"You're wrong," Eva said, still typing furiously as the upload progressed. "Evolution isn't about strength. It's about adaptation. Cooperation. The ability to change." She gestured at his fallen soldiers. "Your rigid ideology made you vulnerable. My mother's inclusive approach made humanity resilient."

An explosion shook the laboratory as Phoenix Order forces breached the rear wall. But instead of the coordinated assault Eva expected, chaos erupted. Some operatives fired on others, shouting about betrayal and contamination. The counter-virus hadn't just affected their bodies—it had shattered their unity.

"Sixty percent uploaded," Eva reported. "Just a few more minutes."

Ben engaged the confused attackers, his tactical superiority evident despite being wounded. But more kept coming, and his ammunition was running low. Eva heard his weapon click empty just as a massive figure emerged through the smoke.

Colonel Marcus Steiner, the Phoenix Order's military commander. Six and a half feet of engineered perfection, enhanced through decades of genetic modification. The counter-virus seemed to have no effect on him.

"Stand down, analyst," Steiner commanded, his voice resonating with authority. "You've lost. Accept it with dignity."

Ben rose slowly, hands empty but stance ready. "I've had worse odds."

Steiner actually smiled. "In Syria? I read the reports. Impressive work. You could have been one of us."

"I've seen what you become." Ben shifted his weight, favoring his unwounded side. "I'll pass."

The two men circled each other while Eva frantically completed her upload. Seventy-five percent. Eighty. Almost there.

Steiner struck first, his enhanced speed blurring the attack. But Ben had fought enhanced opponents before. He redirected the colonel's momentum, using a technique to counter raw power. They crashed into laboratory equipment, sophisticated machinery becoming improvised weapons.

"Ninety percent," Eva announced. "Ben, almost done!"

Steiner had Ben in a chokehold, enhanced muscles slowly crushing his windpipe. Ben's movements grew weaker, his wounds and exhaustion finally overwhelming skill and determination.

"No!" Eva grabbed the nearest object—a microscope—and swung it at Steiner's head. The colonel barely flinched, but it distracted him enough for Ben to break free.

"Upload complete," the terminal announced.

Eva hit the execute button, sending the counter-virus data flooding through maintenance networks worldwide. Within minutes, it would reach thousands of facilities. Within hours, synthesis would begin. The race between virus and cure had truly begun.

Steiner threw Ben across the room and advanced on Eva. "Clever. But we control those facilities. We'll suppress your data, discredit it as terrorism."

"Some of you will," Eva agreed, backing away. "But not all. There are still good people in the world, Colonel. People who'll recognize hope when they see it."

She bumped into something—the facility's main server rack. An idea sparked, desperate but possible.

"You know what your problem is?" Eva asked, fingers finding the emergency shutdown switches. "You think in terms of control. My mother thought in terms of chaos."

She triggered the emergency shutdown. Every system in the facility began failing simultaneously. Lights died, ventilation stopped, and electronic locks disengaged. In the darkness, she heard Steiner curse and Ben move.

Emergency lighting kicked in seconds later, revealing Ben and Steiner grappling on the floor. The colonel's enhancements gave him advantages, but the system shutdown had disabled the facility's oxygen enrichment. At this altitude, even enhanced lungs struggled.

Eva watched the two men fight, knowing she couldn't physically help. But there were other ways to fight. She moved to the chemical storage unit, pulling out compounds she'd worked with for years. Not to create a weapon, but something else.

"Colonel!" she called out. "You're proud of your enhancements, aren't you? Your genetic superiority?"

Steiner glanced at her, maintaining his grip on Ben. "I am what humanity should become."

"Then breathe deep." Eva shattered a container of synthesized pheromones, the same markers she'd embedded in the Phoenix virus to identify targets. But concentrated like this, they'd overwhelm any enhanced olfactory system.

Steiner reeled as the chemical cloud hit him, his enhanced senses becoming a liability. Ben capitalized, driving his knee into the colonel's solar plexus and rolling free.

"You think... this matters?" Steiner gasped, tears streaming from overloaded sensory organs. "The virus spreads. Millions die. Your cure... too slow..."

"Maybe," Eva admitted, helping Ben to his feet. "But it spreads. That's all that matters. Every life saved proves you wrong. Every survivor who helps another negates your philosophy."

The facility shook as explosions cascaded through its lower levels. The system shutdown had triggered catastrophic failures in the containment systems. Soon, the entire complex would be uninhabitable.

"We need to go," Ben urged, leaning heavily on Eva.

She looked back at von Hess, still alive but fading. At Steiner, he was blinded by his own enhancements. At the laboratory where she'd spent her youth dreaming of saving the world.

"The emergency bunker," she decided. "It has independent life support and communication systems. We can monitor the cure's spread, coordinate with anyone who'll listen."

They left the Phoenix Order leaders in the dying facility, making their way through corridors filled with smoke and fleeing personnel. Some tried to stop them, but most were focused on their own survival. The counter-virus had done more than affect their bodies—it had shattered their faith in genetic destiny.

The bunker lay deep beneath the facility, accessed through passages Eva had played in as a child. She sealed the heavy doors behind them, activating systems that would keep them alive while the complex above destroyed itself.

"Now what?" Ben asked, collapsing onto a medical cot.

Eva moved to the communication array, already seeing responses to the data she had uploaded. A laboratory in Mumbai is requesting clarification. A university in São Paulo is beginning synthesis. A hospital in Nairobi is reporting successful trials.

"Now we guide them," she said. "Answer questions. Provide support. Show the world how to save itself."

Ben watched her work, pride evident despite his exhaustion. "Your mother would be amazed."

"She always knew this day might come." Eva pulled up the global infection map, watching as the red zones pulsed and spread. But now there were blue points too—places where the cure was taking hold. "She just made sure we'd be ready."

The hours that followed were a blur of communications, calculations, and careful hope. For every facility that rejected their data, two more accepted it. For every government that suppressed the information, humanitarian organizations spread it wider. The Phoenix Order had underestimated humanity's capacity for cooperation in crisis.

"Eva," Ben said softly, watching the blue zones multiply on their screens. "We did it."

"We started it," she corrected. "Humanity will finish it. Together."

Outside their bunker, the Alpine facility burned. The Phoenix Order's dream of controlled evolution died in flames and chaos. But in laboratories and hospitals around the world, Elena Richter's true legacy lived on—proof that science served best when it served all.

The final approach had begun, not to a world of genetic purity, but to one of shared survival.

The Phoenix had risen and fallen.

Now it was humanity's turn to soar.

# Chapter 32: The Price of Salvation

The emergency bunker's sterile confines couldn't mask the acrid smoke seeping through the ventilation system. Above them, the Alpine facility continued its death throes, explosions cascading through level after level as containment systems failed catastrophically. Eva worked frantically at the communication terminal, coordinating with laboratories worldwide as they attempted to synthesize the counter-virus from her uploaded data.

"Barcelona reports synthesis failure," she announced, reading the incoming messages. "They can't stabilize the protein structures. Tokyo's having similar problems with the RNA sequences."

Ben limped to her side, his wounds hastily bandaged but still seeping blood. "What's wrong?"

Eva pulled up the synthesis reports, her heart sinking as she recognized the pattern. "The data corruption from the emergency upload. Some of the molecular structures didn't transfer completely. They're missing crucial binding sequences."

"Can you resend it?"

"The facility's main transmitters are gone. This bunker only has short-range emergency communications." Eva's fingers flew across the keyboard, trying to reconstruct the missing data from memory. "I need to provide a complete template, a working sample they can reverse-engineer."

Ben's expression darkened as he grasped the meaning. "The only complete sample is what you synthesized in the lab. Which is now dispersed through the facility's air system."

"Not the only sample." Eva moved to the bunker's small medical station, pulling out supplies. "I can synthesize more, but I need to test it first. Ensure it actually works against the Phoenix virus."

"We don't have any of the Phoenix virus here," Ben pointed out.

Eva met his eyes steadily. "Yes, we do."

The realization hit him like a physical blow. "No. Absolutely not. Eva, there has to be another way."

"There isn't." She prepared a syringe with practiced efficiency. "The test sites in Africa are reporting hundreds of deaths per hour. The airport releases are spreading exponentially. Every minute we delay costs lives."

"Then I'll do it. Test it on me."

"You've already been exposed to the virus and the partial antidote. Your system is compromised; the results wouldn't be reliable." Eva extracted a vial from the medical station's secure storage—one of the Phoenix virus samples she'd grabbed during their escape. "It has to be someone with an uncontaminated baseline. It has to be me."

The bunker shuddered as a massive explosion rocked the facility above. Dust rained from the ceiling, and several monitors flickered and died.

"The facility's geothermal plant," Eva noted clinically. "We have maybe an hour before the bunker's independent systems fail."

"Then we run. Get to another facility, find another way—"

"Ben." Eva touched his face gently. "You know, there is no other way. The Phoenix Order's virus is adapting, evolving. Every transmission makes it stronger. If we don't provide a working counter-virus template in the next hour, it'll be too late to stop."

She could see the war in his eyes—the analyst's mind recognizing the logic while the man who loved her raged against it. Finally, his shoulders slumped in defeat.

"How can I help?"

Eva kissed him softly, tasting desperation and determination. "Monitor the synthesis. Upload everything I do. Make sure the world sees exactly how this works."

She moved to the medical chair, setting up recording equipment and diagnostic sensors. Every second of what came next had to be documented, transmitted, and verified. Her death would mean nothing if the cure died with her.

"Phoenix virus, Variant One," she narrated for the recording. "Initial exposure via subcutaneous injection. Estimated time to symptom onset: fifteen minutes. Estimated time to terminal condition: forty-five minutes to one hour."

Her hand was steady as she pressed the syringe to her arm. Ben watched, fists clenched, as she pushed the plunger.

"Injection complete. Beginning counter-virus synthesis using modified protocol seven."

The work was familiar, even as she felt the virus beginning its deadly journey through her bloodstream. Eva had spent years perfecting viral delivery systems. Now she raced to perfect viral destruction, using her own body as the proving ground.

"Munich's attempting synthesis," Ben reported, his voice carefully controlled. "They need clarification on the protein folding sequences."

Eva provided the information while mixing compounds; her hands remained steady, despite knowing what was happening inside her. The Phoenix virus was elegant in its lethality, targeting specific genetic markers to begin cellular breakdown. She could feel it working, a warmth spreading from the injection site.

"First symptoms presenting," she noted clinically. "Elevated temperature, mild disorientation. Increasing synthesis speed to compensate."

Her fingers flew across the equipment, combining elements with precision born from years of practice. The counter-virus took shape in the synthesis chamber—her mother's final gift made manifest through Eva's sacrifice.

That's when the bunker's main door exploded inward.

Wilhelm von Hess stood in the smoke-filled entrance, his elegant suit torn and bloodied but his bearing still aristocratic. Half his face showed burns from the facility's destruction, but his visible eye blazed with fanatic determination.

"Did you think a little fire would stop me?" he asked, stepping over the ruined door. "The Phoenix rises from ashes, Dr. Richter. Always."

Ben moved to intercept, but von Hess was faster. A concealed blade flashed, catching Ben across the ribs. As Ben staggered, von Hess's knee drove into his wounded side, sending him crashing into equipment.

"I've lost my facility, my army, my life's work," von Hess continued conversationally, advancing on Eva. "But I can still ensure you don't live to see your pyrrhic victory."

Eva didn't stop working. The counter-virus was nearly complete, but she needed ten more minutes. Her hands shook now as the Phoenix virus progressed, but she forced them steady through will alone.

"Ben was wrong about you," she said, injecting the synthesized counter-virus into her other arm. "You're not a visionary or a monster. You're just a small man afraid of a changing world."

Von Hess's face contorted with rage. He lunged at her, but Ben intercepted, tackling him away from the synthesis equipment. The two men crashed into the communication array, destroying several monitors.

"Fifteen laboratories now attempting synthesis," Eva reported, her voice growing hoarse. "Sending updated protocols based on counter-virus interaction with the Phoenix variant."

She could feel both viruses battling in her system—the Phoenix virus trying to tear her apart at the cellular level. At the same time, her counter-virus fought to neutralize it. The sensation was like burning from the inside out, every nerve ending screaming as her body became a battlefield.

Ben and von Hess traded brutal blows, their fight destroying much of the bunker's equipment. But Eva noticed Ben targeting specific systems—destroying backups and redundant communications. He was limiting von Hess's options while protecting the primary transmission array.

"Cellular degradation accelerating," Eva noted, coughing. Blood speckled her lips. "But counter-virus is adapting. It's working."

She uploaded real-time data from her body's battle, providing laboratories worldwide with a precise template for their research. São Paulo reported successful synthesis. Mumbai confirmed positive results. The blue dots on her map multiplied as humanity's scientific community united against extinction.

Von Hess gained the upper hand, his blade finding Ben's shoulder. As Ben fell, von Hess turned toward Eva with murderous intent.

"Your mother thought she was so clever," he snarled, advancing despite his injuries. "But I understood her work better than she did. The counter-virus you're so proud of? It has a fatal flaw."

Eva's vision blurred as the Phoenix virus ravaged her system. "What flaw?"

"It requires a living host to properly calibrate." Von Hess smiled through bloodied teeth. "The synthesized versions will be inferior, degraded. They'll slow the Phoenix virus but not stop it. Unless..."

"Unless they have a perfect template," Eva finished, understanding flooding through her. "A living sample showing exactly how the counter-virus defeats the Phoenix strain."

"Which dies with you." Von Hess raised his blade.

Ben's gunshot caught von Hess in the chest, spinning him around. The Phoenix Order's leader looked down at the spreading crimson with genuine surprise.

"Impossible," he whispered. "You were disarmed..."

"I was," Ben agreed, holding Eva's emergency sidearm. "But Eva wasn't."

Von Hess fell to his knees, then forward, his blood pooling on the bunker floor. His final words were barely audible: "The strong... survive..."

"No," Eva corrected, her voice fading. "The connected survive. The ones who help each other."

Ben rushed to her side as she collapsed from the chair. The diagnostic readings showed the war inside her reaching its climax—cellular breakdown accelerating even as the counter-virus adapted to fight it.

"The upload," she gasped. "Is it complete?"

Ben checked the displays through tears. "Forty-seven laboratories have successfully synthesized. More coming online every minute. Eva, it's working. The cure is spreading."

She smiled, tasting copper and victory. "Then it's enough. Tell them... tell them to watch the protein sequences in my blood. The counter-virus is learning, adapting. They need to see..."

Her body convulsed as the Phoenix virus made a final assault. But the counter-virus had learned too much, adapted too well. It struck back with precision that would have made Elena Richter proud, neutralizing the weapon at its source.

"Eva, stay with me," Ben pleaded, cradling her as she shook. "The cure's working. You're going to be fine."

But she knew better. The battle inside her had taken too much. The counter-virus would win, but her body was too damaged to survive the victory.

"Record everything," she whispered. "Every second. Every change. Let them see how it works."

Ben activated every remaining sensor, documenting her body's final moments as the counter-virus achieved complete victory over the Phoenix strain. The data streamed out to waiting laboratories, providing the perfect template they needed.

"Ninety-three facilities confirming synthesis," Ben reported, his voice breaking. "WHO is coordinating global distribution. Major cities are already producing doses. You did it, Eva. You saved them all."

"We did it." Her hand found his, weak but warm. "You, me, mother. Even father, in his twisted way. All of us together."

The bunker's emergency lighting flickered as power systems failed. However, the communication array, powered by backup batteries, continued to transmit Eva's final gift to humanity.

"I wish we had more time," she whispered.

"We have forever," Ben said fiercely. "Every life saved carries your memory. Every child who grows up in the world you preserved. You're not dying, Eva. You're becoming eternal."

She managed one last smile. "That's beautiful. But still a lie."

"The best kind," he agreed, holding her as the light faded from her eyes.

The diagnostic systems recorded the exact moment of her death—and the exact moment the counter-virus achieved perfect synthesis in her blood. The template transmitted to thousands of facilities, carrying with it the solution Elena Richter had hidden in a lullaby and Eva had proven with her life.

Ben held her body as the bunker went dark around them, lit only by the communication array's status lights. Each light represented a laboratory that had successfully created the cure. A constellation of hope born from sacrifice.

"Processing complete," the system announced. "Counter-virus template distributed to all connected facilities. Estimated global synthesis capacity: ten million doses per day. Projection: Phoenix virus neutralization within thirty to forty-five days."

Ben kissed Eva's forehead gently. "You heard that? Thirty days. You gave humanity thirty days to save itself. That's all we ever needed."

The bunker fell silent except for the hum of transmitting data. Outside, the Alpine facility burned itself out, taking the Phoenix Order's dreams with it. But in Eva Richter's blood, in the template born from her sacrifice, humanity's future lived on.

The synthesis was complete.

The Phoenix virus would claim millions before the cure reached them all. But it would not claim humanity. A young woman's courage, a mother's paranoid brilliance, and the collective effort of scientists worldwide would ensure that.

Ben stayed with Eva until rescue teams found them twelve hours later. He refused to leave her body, insisting she be treated with the honor due humanity's savior. The world would know her name, her sacrifice, her victory.

But first, the world had to survive. And thanks to Eva Richter, it would.

The synthesis had required everything—brilliance, determination, sacrifice, and love.

She had given them all.

# Chapter 33: The Phoenix Burns

The rescue team's lights pierced the bunker's darkness twelve hours after Eva's death. Ben hadn't moved from her side, her body cradled in his arms as if his embrace could somehow restore the life that had saved millions. The communication array continued its work, each pulse of data carrying her legacy further across the globe.

"Sir, we need to get you medical attention," the lead rescuer said gently. A Swiss military medic, her uniform bearing the insignia of the international response team. "Your wounds—"

"Not yet." Ben's voice was hoarse from grief and smoke. "She stays with me."

The medic exchanged glances with her team but nodded. They'd been briefed on what had happened here, what Eva Richter had done. Heroes deserved their vigils.

As they prepared a stretcher for Eva's body, Ben noticed movement in the bunker's shadows. A figure stumbled forward, clothes torn and burned, one arm hanging uselessly. Klaus Richter had somehow managed to survive the facility's destruction.

"My daughter," Klaus whispered, seeing Eva's still form. "My brilliant, stubborn daughter."

Ben's hand moved to his weapon, but Klaus collapsed to his knees before he could draw it. The architect of genocide looked broken, aged decades in hours.

"I helped create the counter-virus," Klaus said, his words tumbling out in desperate confession. "When I realized what Eva was doing, I... I added my own research. Transmitted it from the facility's backup servers before they burned. The protein sequences she was missing—I provided them. It was all I could do."

"You murdered her mother," Ben growled. "You created a weapon to kill billions. You hunted your own daughter."

"Yes." Klaus didn't deny it, couldn't deny it. "I became a monster in pursuit of perfection. But Eva... she showed me what true evolution looks like. Not through selection, but through sacrifice. Through love."

The rescue team secured Eva's body with infinite care, preparing to carry her from this tomb of ambitions. Klaus reached out with his good hand, not quite touching her.

"May I... may I say goodbye?"

Ben wanted to refuse. This man had caused so much death, so much suffering. But Eva had believed in redemption, even at the end. Slowly, he nodded.

Klaus bent close to his daughter's ear, whispering in German. Ben caught fragments—apologies, regrets, and strangely, pride. When Klaus straightened, tears had carved channels through the soot on his face.

"The Phoenix Order will try to regroup," Klaus said, his voice steadier now. "Von Hess was just one leader. There are others, hidden in governments and corporations worldwide. I know them all."

"Then you'll tell us everything," the rescue team leader said, producing restraints. "Klaus Richter, you're under arrest for crimes against humanity."

Klaus submitted without resistance, but his eyes stayed on Ben. "She loved you. It was in her voice every time she spoke your name. That love gave her the strength to save everyone. Remember that when the grief threatens to consume you."

They carried Eva's body from the bunker with the reverence reserved for heads of state. Ben followed, supported by medics who insisted on treating his wounds. As they emerged into daylight, he saw the full scope of the facility's destruction. The Phoenix Order's stronghold had become a crematorium, black smoke still rising from its ruins.

But his communication device was buzzing with updates from around the world. The counter-virus was working.

"Sir," a young officer approached. "You need to see this."

She handed him a tablet showing global news feeds. In every major city, laboratories and hospitals were producing Eva's cure. Distribution had begun in the hardest-hit areas. The death tolls were staggering—thousands per hour—but the rate was slowing. The Phoenix virus had met its match.

More importantly, humanity was responding as Eva had predicted. Countries that had been enemies shared research freely. Pharmaceutical companies opened their patents. Doctors worked without regard for borders or politics. Faced with extinction, humanity had chosen cooperation over competition.

"She did this," Ben murmured, watching distribution centers operate with military efficiency. "She knew we could unite when it mattered."

"Mr. Carter," a familiar voice called. Sarah Chen's replacement at the CIA, Director Marcus Webb, approached with a small army of intelligence officials. "We need to debrief you immediately. The Phoenix Order—"

"Can wait," Ben interrupted. "I'm accompanying her body."

"Sir, with respect, time is critical. Phoenix Order cells are activating worldwide. Some are trying to sabotage the distribution of the cure. Others are destroying evidence. We need your tactical assessment—"

Ben rounded on him with such fury that security agents reached for weapons. "She gave everything. Everything! The least we can do is see her home with dignity."

Webb stepped back, hands raised placatingly. "Of course. But perhaps you can help us while—" He gestured to the tablet.

Ben looked at the screen, which showed Phoenix Order facilities across the globe. Some were burning like the Alpine complex. Others were under siege by military forces. But in a disturbing number, the personnel had simply vanished, leaving behind empty buildings and scrubbed servers.

"They're running," Ben realized. "Scattering to regroup later."

"Our thoughts exactly. Your insight could prevent—"

"Director Webb." The voice was quiet but carried absolute authority. The Swiss Federal Councilor for Defense had arrived, accompanied by representatives from the United Nations. "Mr. Carter has earned the right to grieve. We'll handle the immediate crisis."

The Councilor turned to Ben. "Dr. Richter will receive a state funeral. The world needs to mourn its savior properly. You have my word she'll be honored as she deserves."

Ben nodded, not trusting his voice. They loaded Eva's body into a medical transport with infinite care. He climbed in beside her, ignoring his own injuries. Through the vehicle's windows, he could see the mountains she'd loved as a child, where she'd returned to face her father's madness.

The transport's communication system came alive with reports as they descended toward Zurich. New York had established distribution centers in all five boroughs. London's hospitals were reporting successful treatments. Tokyo's efficient response resulted in limited casualties, numbering in the hundreds rather than the thousands.

But there were darker reports too. Phoenix Order loyalists had bombed a synthesis facility in Cairo. Armed groups were trying to prevent distribution in regions they controlled, claiming the cure was a Western plot. Misinformation spread almost as fast as the virus itself.

"She knew this would happen," Ben said to no one in particular. "That's why she made the synthesis process so simple, why she spread it so widely. Too many sources to stop them all."

His secure phone rang. The caller ID displayed the name of the President of the United States.

"Mr. Carter, I'm calling to express our nation's profound gratitude—"

"Thank Dr. Richter," Ben interrupted. "Thank her mother. Thank everyone who chooses to help rather than hoard the cure. I just held her hand while she died."

"I understand you're grieving—"

Ben ended the call. World leaders could wait. Eva deserved better than being reduced to a political talking point while her body was still warm.

They reached Zurich as the sun set, painting the sky in shades of flame. Fitting, Ben thought, for the day the Phoenix Order burned. Their careful plans, their hidden members, their dreams of controlled evolution—all reduced to ash by one woman's sacrifice.

The hospital had prepared a private room where Eva's body could rest before the funeral. Ben stood guard, turning away everyone except the doctors who needed to examine her. Even in death, she was teaching them. Her blood samples showed exactly how the counter-virus worked, providing data that would save millions more.

"Mr. Carter?" An elderly woman entered, her face familiar from scientific journals. Dr. Marie Brunner, one of Elena Richter's contemporaries. "I worked with Eva's mother. May I?"

Ben stepped aside. Dr. Brunner approached Eva's body with tears streaming down her weathered face.

"Elena would be so proud," she whispered. "She always said Eva would change the world. Just... not like this." She turned to Ben. "I have something you should know. Elena didn't work alone on the original virus. There were three of us—Elena, I, and David Huang. We thought we were creating medical miracles."

"You couldn't have known what it would become," Ben offered.

"Couldn't we?" Dr. Brunner's laugh was bitter. "The funding was too good, the security too tight. We suspected, but we wanted to believe in the work more than we wanted to see the truth. Elena was the only one brave enough to look behind the curtain."

She pulled out a worn notebook. "After Elena died, David and I went into hiding. We've spent fifteen years trying to understand what she discovered, searching for the counter-virus we knew she must have created. We were so close, but Eva... Eva found it first."

"She found it in herself," Ben said. "Her mother hid it in the one place Klaus would never think to look—his own daughter."

Dr. Brunner smiled through her tears. "Brilliant. Elena always was ten steps ahead of everyone." She placed a gentle hand on Eva's forehead. "Rest now, dear girl. You've earned peace."

As she left, Ben's phone buzzed with encrypted messages. Intelligence agencies worldwide were coordinating the hunt for fleeing Phoenix Order members. Klaus's interrogation had already yielded dozens of names and locations. The organization that had planned to reshape humanity was being systematically dismantled.

But Ben knew better than to think it was over. Ideologies didn't die with their leaders. Somewhere, Phoenix Order survivors were regrouping, planning, waiting for the world to forget. They'd return eventually, in new forms with new justifications for the same old hatred.

"Not on my watch," he murmured to Eva's still form. "I promise you that."

A commotion outside drew his attention. Through the window, he could see crowds gathering despite the late hour. They carried candles, flowers, and signs reading "Thank You" in a dozen languages. Word had spread about where Eva's body rested. The city of Zurich was coming to pay its respects.

Hospital security attempted to maintain order, but the crowd continued to grow larger by the minute. These weren't dignitaries or officials. They were ordinary people whose lives had been saved by Eva's sacrifice. Parents with children, elderly couples, young students—humanity in all its chaotic diversity.

"Sir," a security officer entered. "We can't hold them back much longer. Should we move the body?"

Ben looked at the gathering crowd, then at Eva. She'd believed in people, in their capacity for good despite evidence to the contrary. These mourners were proving her right.

"Let them in," he decided. "Ten at a time. Give them a chance to say goodbye."

For hours, they came. Each group was different—different in age, race, and nationality—precisely the diversity the Phoenix Order had tried to eliminate. They left flowers until the room overflowed with them. They whispered thanks in languages Ben didn't recognize. Some prayed, others simply stood in silent witness.

A young girl, perhaps seven years old, tugged at Ben's sleeve. "Is she really the lady who made the medicine?"

"Yes," Ben managed through his tight throat. "She made the medicine that's saving everyone."

The girl nodded solemnly and placed a hand-drawn picture on the growing memorial. It showed a figure in a lab coat with angel wings, surrounded by hearts. "My teacher says she's a hero like in stories. But better because she's real."

"Much better," Ben agreed.

By dawn, thousands had paid their respects. The hospital room had become a shrine to humanity's salvation. And through it all, the communication networks buzzed with success stories. Mumbai reported zero new infections. São Paulo's hospitals were releasing recovered patients. Beijing announced it had surplus doses to share with rural regions.

The Phoenix had burned, but from its ashes rose something unexpected—not a new world order of genetic purity, but a reminder of what humanity could accomplish together. Eva had given them more than a cure. She'd given them proof that cooperation could triumph over division, that sacrifice could defeat selfishness.

Director Webb returned as the sun rose over Zurich. "Mr. Carter, I apologize for yesterday's insensitivity. However, we've tracked significant movement of the Phoenix Order toward South America. We could use—"

"I'll help," Ben interrupted, his decision crystallizing. "But first, she gets the funeral she deserves. Then I hunt down every last one of them. That's the deal."

Webb nodded. "Understood. The President has authorized the provision of any resources you may need. The Phoenix Order ends with us."

Ben looked at Eva one last time. Her face was peaceful, almost smiling. She'd known the cost and paid it willingly. Now it was his turn to ensure her sacrifice meant something permanent.

"Burn them all," he said quietly. "Every cell, every supporter, every trace of their poisonous ideology. The Phoenix rises from ashes, but some things need to stay dead."

The Phoenix Order had dreamed of fire that would purify humanity.

Instead, they'd created the flame that would consume them entirely.

And Ben Carter would tend that fire until nothing remained but justice and ash.

# Chapter 34: The Last Goodbye

The Cathedral of St. Pierre in Geneva had witnessed history—Calvin's sermons, diplomatic marriages, and treaties that shaped nations. But nothing compared to the gathering assembling within its ancient walls for Eva Richter's funeral. World leaders sat beside laboratory technicians. Survivors of the Phoenix virus filled pews with those still fighting it. Humanity's diversity is packed into stone and stained glass, united in grief and gratitude.

Ben stood in the anteroom, adjusting the black suit someone had provided. His wounds were now properly treated, held together by surgical skill rather than desperate bandaging. But the deepest wound was beyond any doctor's ability to heal.

"Mr. Carter," President Harrison of the United States approached, flanked by security. "I wanted to personally—"

"Thank you, Madam President," Ben cut her off politely but firmly. "But please, save the speeches for the service. Eva deserves better than political platitudes in private."

The President nodded, surprisingly. "You're right. I just..." She paused, the leader of the free world momentarily at a loss. "My granddaughter is alive because of her. I wanted you to know that the person thanking you isn't just the President. It's a grandmother."

She moved on, leaving Ben with the weight of personal gratitude that had been pouring in for three days. Every world leader, every scientist, everyone who'd lost someone or saved someone wanted him to know what Eva meant to them. As if he could somehow pass the messages along.

"Ben." Sarah Chen entered—not her replacement, but Sarah herself, alive despite the assassination attempt he'd witnessed. Her arm was in a sling, but her eyes were clear. "I'm sorry. I should have contacted you sooner, but the agency kept my survival classified while we hunted the mole who leaked my location."

"Sarah." Relief flooded through him. One less death on their conscience. "Eva would be glad you're alive."

"She saved us all." Sarah's professional composure cracked slightly. "I've been coordinating the intelligence response. The Phoenix Order is shattered, but fragments are trying to regroup. We need—" She stopped herself. "Later. Today is for her."

Klaus Richter entered under heavy guard, shackled but dressed in a simple black suit. The authorities had debated allowing him to attend, but Ben had insisted. Eva would have wanted her father there, despite everything.

"Thank you," Klaus said simply. The past three days of interrogation had aged him further. Every name he'd revealed, every secret he'd exposed, carved new lines in his face. "She deserved better than a monster for a father."

"She deserved the father you could have been," Ben replied. "The one who chose love over ideology."

"Yes." Klaus stared at his shackled hands. "I've given them everything—names, locations, protocols. Forty-seven Phoenix Order cells have been raided based on my information. But it's not enough. It will never be enough."

The cathedral bells began to toll, their bronze voices carrying across Geneva and beyond. It was time.

The walk to Eva's casket seemed endless. She lay surrounded by flowers from every corner of the earth—orchids from Thailand, roses from Ecuador, lilies from Holland. Her face was peaceful, the mortician's art hiding the ravages of her final battle. She wore a simple blue dress, the color she'd loved as a child, building sandcastles on the Baltic shore.

Ben took his place in the front row, Klaus heavily guarded beside him. The cathedral filled to capacity and beyond, with thousands more gathering outside to watch on screens. The world had stopped to say goodbye.

The service began with a children's choir singing Eva's mother's lullaby—the song that had hidden humanity's salvation. Their pure voices filled the cathedral with the melody of redemption, and Ben saw hardened diplomats weeping openly.

Chancellor Müller of Germany spoke first, her voice carrying the weight of her nation's complex history. "Eva Richter proved that the greatest strength isn't in selecting who survives, but in ensuring everyone has the chance to. She chose the harder path—saving those who would have killed her, healing those marked for death. That is true evolution."

Dr. Chen Wei from Beijing followed, representing the global scientific community. "In seventy-two hours, we've produced over thirty million doses of the counter-virus. Distribution reaches the most remote villages and the largest cities equally. Dr. Richter's synthesis protocol was elegant in its simplicity—she wanted everyone to be able to create hope."

Survivors spoke next. A teacher from Lagos who had been among the first infected at the testing sites. A child from Cairo who'd received the cure just hours before organ failure. An elderly man from New York who'd lost his wife to the virus but lived to see their grandchildren saved. Each story was a testament to Eva's sacrifice.

President Harrison took the podium, her political polish set aside. "We gather today not just to mourn a hero, but to commit ourselves to the world she died creating. A world where cooperation trumps competition, where diversity is strength, where science serves all humanity. The Phoenix Order tried to burn that world down. Eva Richter ensured it would rise from the ashes."

Ben had been asked to speak but declined. His goodbye to Eva was private, whispered in the darkness of a bunker, sealed with a final kiss. Instead, he'd asked that her work speak for her.

The cathedral's screens lit up with images from around the world. Synthesis laboratories are working around the clock. Distribution centers operate with military efficiency. Recovered patients leaving hospitals. Children playing in streets that should have been death zones. Life continues because one woman refused to let it end.

But it was the final speaker who brought the most profound silence. Klaus Richter, escorted to the podium in chains, faced the assembly. The architect of genocide, looking at those he would have murdered.

"My daughter understood something I spent a lifetime denying," he began, his voice barely above a whisper. "That humanity's strength comes not from genetic purity but from genetic diversity. Not from selecting the strong but from protecting the vulnerable. I created a weapon to reshape the world in my twisted image. She transformed it into a tool that preserved the world in all its chaotic beauty."

He paused, gathering himself. "I've provided the authorities with everything needed to dismantle the Phoenix Order. But information isn't absolution. I know where I'm going after this—to a cell where I'll spend whatever time remains contemplating the monstrosity of my actions. It's less than I deserve."

His eyes found Ben. "But before I go, I need to say what I couldn't while she lived. Eva, my brilliant, compassionate daughter, you were everything I claimed to be building toward. The next step in human evolution. Not through engineered genetics but through chosen sacrifice. You proved that love is the most powerful force in nature. Forgive me for taking so long to see it."

Klaus broke down completely, his sobs echoing in the cathedral's vastness. Guards moved to escort him away, but Ben raised a hand. Let him grieve. Let the monster remember his humanity in the face of what he'd lost.

As Klaus was led back to his seat, the ceremony drew to a close. Religious leaders from every major faith offered prayers in their traditions. Muslims, Christians, Jews, Hindus, Buddhists—all united in blessing Eva's journey beyond. The Phoenix Order had tried to divide humanity along genetic lines. Eva's funeral proved those divisions were illusions.

The pallbearers stepped forward—Ben, Director Webb, Dr. Chen Wei, Dr. Marie Brunner, and two survivors chosen to represent the millions saved. They lifted Eva's casket with reverence, beginning the slow processional out of the cathedral.

As they emerged into daylight, Ben gasped. The streets of Geneva were lined with people as far as he could see. They stood in silent tribute, many holding candles despite the afternoon sun. As the casket passed, they began to applaud—not the celebration of crowds, but the deep appreciation of the saved for their savior.

The journey to the cemetery continued through this corridor of humanity. Ben saw tears on countless faces, hands reaching out to touch the hearse as it passed. These weren't world leaders or scientists. They were ordinary people whose ordinary lives continued because of Eva's extraordinary sacrifice.

At the graveside, a smaller gathering of those who'd known Eva personally. Her childhood teachers, university colleagues, friends from her years of research. Each dropped a flower into the grave, a personal goodbye to the woman they'd known before she became humanity's savior.

Ben waited until the others had paid their respects. Then, alone beside the open grave, he pulled out a small box. Inside were two rings—the ones they'd never had a chance to exchange, purchased during a moment of hope between disasters.

"We never talked about marriage," he said quietly. "Too busy saving the world. But I bought these in Prague, that night after we escaped Roth's betrayal. You kissed me on that rooftop, and I knew. Even with the world ending, even with impossible odds, I knew I wanted whatever time we had to be together."

He placed one ring on Eva's finger, the gold bright against her pale skin. The other he pocketed next to his heart.

"The world's safe now. The Phoenix Order is burning itself out. The cure spreads faster than their virus ever could. You did it, Eva. You saved everyone." His voice cracked. "I just wish you'd saved yourself too."

The gravediggers waited patiently as Ben said his final goodbye. He kissed her forehead one last time, tasting formaldehyde and flowers instead of life and love. Then he stepped back, nodding for them to continue.

As earth covered Eva Richter, Ben felt something shift in the air. Not ending but transformation. She was gone, but her work lived on in every dose synthesized, every life saved, every choice to help rather than harm.

"Mr. Carter," Director Webb approached carefully. "I'm sorry to intrude, but we've just received intelligence. A significant Phoenix Order cell is planning to sabotage cure distribution in South America. We could use your expertise."

Ben looked at the fresh grave, then at the Director. Eva had given everything to stop the Phoenix Order. Could he do less?

"Brief me on the way," he said, turning from the cemetery. "She wouldn't want me standing here when there's work to be done."

As they walked toward the waiting vehicles, Ben noticed Klaus being loaded into a secure transport. Their eyes met briefly. In that moment, Ben saw not the architect of genocide but a broken father who'd realized too late what truly mattered.

"Take care of her legacy," Klaus called out. "Don't let them return. Don't let her death be for nothing."

"Never," Ben promised.

The transport doors closed, taking Klaus to justice and penance. Ben climbed into his own vehicle, already reading intelligence reports on the Phoenix Order remnants. The organization that had planned so carefully for victory had never planned for defeat. They were scattered, desperate, dangerous in their death throes.

Good. Ben preferred hunting desperate prey. They made mistakes, took risks, exposed themselves. And he would be there, ensuring Eva's sacrifice remained permanent.

As Geneva fell behind them, Ben touched the ring in his pocket. Eva was gone, but their war continued. Every Phoenix Order cell he destroyed, every cure distribution he protected, every life he saved would be a love letter to her memory.

The last goodbye had been said. Now came the long work of ensuring it meant something.

In laboratories worldwide, Eva's cure continued its work. In intelligence centers, her father's testimony dismantled terrorist networks. In Ben's heart, her memory drove him forward.

The Phoenix had tried to burn the world. Eva had transformed that fire into light.

And Ben would tend that light until his own last goodbye came.

The world would remember Eva Richter as the woman who saved humanity.

Ben would remember her as the woman who saved him.

Both memories would endure.

# Chapter 35: The Aftermath

The International Criminal Court in The Hague had never seen security like this. Snipers on every rooftop, armored vehicles at each intersection, and screening procedures that made airport security look casual. Three weeks after Eva's funeral, the world was taking no chances with the Phoenix Order trial of the century.

Ben Carter adjusted his tie in the witness waiting room, his reflection showing a man transformed by loss and purpose. The wounds had healed, leaving scars that mapped his journey from academic to warrior to guardian of Eva's legacy. In those three weeks, he'd helped coordinate seventeen raids across four continents. Each Phoenix Order cell destroyed was another promise kept.

"Mr. Carter," the court officer appeared at the door. "They're ready for you."

The walk to the witness stand felt like crossing into history. The courtroom gallery was packed with journalists, survivors, and families of the dead. Behind reinforced glass sat forty-three Phoenix Order members captured in the global sweep—scientists, military officers, government officials who'd hidden their genocidal allegiance behind respectable facades.

Klaus Richter sat in the front row of defendants, isolated even among his former allies. His testimony had condemned them all, and they knew it. The looks they shot him promised vengeance that prison walls would hopefully prevent.

"State your name for the record," the prosecutor began.

"Benjamin Samuel Carter. Former CIA analyst, current special advisor to the Global Health Security Coalition."

"Mr. Carter, you were with Dr. Eva Richter when she synthesized the counter-virus that saved humanity. Can you describe those final hours?"

Ben's testimony was clinical and precise. He detailed their infiltration of the Alpine facility, the discovery of the Phoenix Order's true scope, and Eva's realization about her mother's hidden legacy. But when he reached the point of Eva's decision to test the counter-virus on herself, his voice faltered.

"She knew it would kill her," he said, meeting the eyes of the defendants. "The synthesis process required a living template, someone to prove the cure worked by surviving the virus. But her body couldn't withstand the battle between the pathogen and the cure. She made that choice knowing the cost."

"Objection," one of the defense attorneys interrupted. "The witness is speculating about Dr. Richter's state of mind."

"I'm reporting her exact words," Ben replied before the judge could rule. "She said, 'Every life saved proves you wrong. Every survivor who helps another negates your philosophy.' She died believing humanity would choose cooperation over your client's vision of selection."

The prosecutor guided him through the technical details—how the counter-virus worked, how quickly it spread, the current global statistics. One hundred forty-seven million doses distributed. Sixty-three countries now manufacture the cure. Death toll stabilizing at three million, horrific but far short of the billions the Phoenix Order had planned.

"The defense may cross-examine," the judge announced.

The lead defense attorney, a sharp-featured woman who had made her reputation defending war criminals, approached with predatory ease.

"Mr. Carter, isn't it true that you've been conducting extrajudicial operations against suspected Phoenix Order members?"

"I've been assisting law enforcement in locating fugitives," Ben replied evenly.

"Assisting?" She produced a tablet showing security footage. "This is you in São Paulo two weeks ago. You're kicking down the door of a medical facility. Three men were hospitalized after this 'assistance.'"

"Those men were preparing to poison the city's water supply with a modified version of the Phoenix virus. The authorities were twenty minutes away. I made a judgment call."

"You made yourself judge, jury, and executioner."

"I made myself a guardian of the cure Dr. Richter died creating." Ben's voice hardened. "Every Phoenix Order member still free is a threat to that legacy. So yes, I hunt them. Yes, I stop them. And yes, I'll continue until every last one faces justice."

The defense attorney started to object, but Klaus Richter suddenly stood in the defendant's box.

"May I address this?" he asked.

The judge looked startled. "Mr. Richter, you have your own counsel—"

"Those whom I've instructed to remain silent. I've pleaded guilty to all charges. But I can speak to Mr. Carter's actions." Klaus turned to face Ben directly. "You want to paint him as a vigilante? He is. Thank God. My former colleagues aren't misguided idealists—they're fanatics who'll kill millions more given the chance. Mr. Carter understands what my daughter understood: some threats require immediate action."

"Mr. Richter," his own attorney hissed, "please sit down."

But Klaus continued. "I've provided intelligence on every cell, every member, every resource. Yet three facilities producing modified viruses were missed. Who found them? Who stopped releases that would have killed thousands? Not the authorities following proper procedures. Him. The man my daughter loved. The man honoring her sacrifice with action while you debate semantics."

The judge gaveled for order as the gallery erupted. Several defendants tried to shout over Klaus, calling him a traitor and worse. Security moved to restore calm, but the damage was done. Klaus had validated Ben's methods before the world.

"Mr. Richter will be removed if there are further outbursts," the judge warned. "Mr. Carter, please continue your testimony."

Ben spent four hours on the stand, detailing every aspect of the Phoenix Order's operations he'd witnessed or uncovered. He named names, described facilities, and explained the twisted ideology that justified genocide as evolution.

By the end, several defendants had paled visibly, realizing how thoroughly their secrets were exposed.

But it was his final statement that resonated beyond the courtroom.

"The Phoenix Order believed in genetic destiny," Ben said, addressing not just the court but the cameras broadcasting globally. "They thought DNA determined worth, that selection required destruction. Dr. Eva Richter proved them wrong with her last breath. She showed that our choices, not our genes, define us. Every person who helps distribute the cure, who cares for the sick, who rebuilds after tragedy—they're evolution in action. Not through death, but through life. Not through selection, but through connection."

As he stepped down from the witness stand, Ben noticed someone in the gallery—Dr. Marie Brunner, Eva's mother's colleague. She nodded slightly, tears on her weathered face. The testimony had honored both mother and daughter.

Outside the courthouse, the media circus was overwhelming. Ben pushed through with security's help, ignoring shouted questions about his "vigilante actions" and "plans for the Phoenix Order remnants." He had work to do.

His secure phone buzzed with a priority alert. Director Webb's face appeared on the encrypted connection.

"Ben, we have a problem. Satellite surveillance shows unusual activity at a decommissioned Soviet bioweapons facility in Kazakhstan. Heat signatures suggest it's operational."

"Phoenix Order?"

"Unknown, but the timing's suspicious. Local assets report trucks arriving at night, personnel in hazmat gear. Could be a new synthesis attempt."

Ben was already moving toward his vehicle. "Send me everything. I'll take a team—"

"Ben." Webb's tone stopped him. "This isn't São Paulo. If the Phoenix Order has taken over a former Soviet facility, it likely has military-grade defenses. You can't kick down this door."

"Then what do you suggest?"

"Let us handle it. Military strike, full containment protocols. You've done enough—"

"Have I?" Ben looked back at the courthouse where humanity's would-be extinctionists faced justice. "Eva didn't stop at 'enough.' Neither do I."

"She also died," Webb said gently. "Is that what she'd want for you?"

The question hung between them. In the three weeks since Eva's death, Ben had barely slept, pushing himself from one operation to the next. Each success felt like a conversation with her, a way to feel connected across the void. Stopping meant confronting the silence where her voice should be.

"She'd want the job finished," Ben finally answered. "Send me the intel. I'll coordinate with local forces."

He ended the call before Webb could argue further. The work continued, had to continue. Every Phoenix Order member captured or killed was another step toward the world Eva had envisioned—one where cooperation trumped competition, where diversity was strength.

His driver, a former Special Forces operator assigned by the Coalition, glanced in the rearview mirror. "Kazakhstan, sir?"

"Eventually. First, Berlin. I need to pick something up."

Four hours later, Ben stood in his old university office. Dust motes danced in afternoon sunlight, highlighting the life he'd abandoned when Eva crashed into his world. His books remained on their shelves, his papers stacked neatly on the desk. A museum of the man he'd been before purpose found him again.

He retrieved what he'd come for—a locked drawer containing his operational files from CIA days. If he was going to take on a militarized facility, he needed more than determination. He needed the skills he'd buried after Syria, the tactical knowledge that had made him valuable before conscience made him a liability.

Among the files, he found something unexpected. A photo from a joint operation in Turkey, him standing beside Sarah Chen and other assets. They looked so young, so certain of their mission's righteousness. Before the wedding party. Before forty-seven civilians died from his analysis.

"You can't hide from the world forever," Sarah had told him in that Berlin café, minutes before her staged death began this journey. She'd been right. Eva had proved it. The world needed people willing to act, even when action carried terrible costs.

Ben's phone rang. Dr. Brunner's name appeared on the screen.

"Mr. Carter, I'm sorry to bother you, but something's happened. My laboratory was broken into last night. Nothing taken, but someone accessed Eva's research files."

"Which files?"

"Everything related to the counter-virus synthesis. They copied it all." She paused. "Ben, only someone with intimate knowledge of our security would know those files existed. This was an inside job."

The implications chilled him. Someone in their own ranks was gathering Eva's research. Not to destroy it—the cure was already global. Which meant they wanted to modify it, perhaps create a new weapon from Eva's salvation.

"Double your security," Ben instructed. "I'll have Coalition forces provide protection. And Doctor? Back up everything to multiple secure locations. We can't let them corrupt her work."

He ended the call and immediately contacted Webb. The Phoenix Order was more than scattered fanatics—they had sleeper agents still hidden, still plotting. The trial in The Hague was just the visible portion. The real war continued in the shadows.

As his plane lifted off for Kazakhstan hours later, Ben reviewed the satellite intelligence. The facility was a Cold War relic, officially decommissioned following the collapse of the Soviet Union. However, someone had reactivated it, and the heat signatures indicated major biological activity was underway.

His team was already assembled—operators from various nations' special forces, united under the Coalition's banner. They knew the stakes. A new bioweapon could undo everything Eva had achieved. Her cure had saved the world once. They couldn't count on lightning striking twice.

"Insertion in six hours," the team leader reported. "Local assets confirm increased security, military-grade weapons. This won't be subtle."

"Good," Ben replied, checking his equipment. "I'm tired of subtle."

He thought of Eva as they flew through darkening skies. She'd believed in redemption, in the possibility of change. But she'd also understood that some threats required decisive action. The Phoenix Order had chosen extinction over evolution. They'd made themselves humanity's enemy.

In The Hague, the trial continued. Klaus Richter and forty-two others faced justice through proper channels. But in hidden facilities and shadowed corners, other Phoenix Order members worked to resurrect their dream of selective genocide.

Ben touched the ring in his pocket—his talisman, his reminder. Eva had given her life to save everyone. He'd spend his ensuring that the sacrifice remained permanent.

The world had survived the Phoenix Order's first assault. But like their namesake, they sought to rise from apparent death. Ben's job was to ensure those ashes stayed cold, that particular fire never rekindled.

Kazakhstan awaited. Another facility to destroy, another cell to eliminate. The work would continue until every last trace of the Phoenix Order was expunged from the earth.

Eva had saved the world with synthesis and sacrifice.

Ben would keep it safe with determination and, when necessary, violence.

Different methods, same goal—ensuring humanity's future remained unselected, uncontrolled, beautifully diverse in its chaotic potential.

The aftermath wasn't an ending.

It was a beginning.

# Chapter 36: The Reckoning

Three months had passed since Eva Richter's death, and the world had transformed in ways both visible and hidden. The Geneva Tribunal's main chamber hummed with quiet efficiency as Ben Carter took his seat at the witness table for the forty-seventh time. Each testimony peeled back another layer of the Phoenix Order's conspiracy, revealing depths of corruption that even seasoned investigators found shocking.

"Mr. Carter," the lead prosecutor began, her voice carrying the weight of accumulated horrors. "Today, we need to address the Phoenix Order's infiltration of the World Health Organization. Your recent operation in Mumbai uncovered disturbing evidence."

Ben activated his tablet, displaying classified documents on the courtroom screens. "Three weeks ago, we raided a WHO subsidiary laboratory that was supposedly distributing the counter-virus. Instead, we found they were collecting genetic samples from everyone who received the cure."

Gasps rippled through the gallery. Even after months of revelations, the Phoenix Order's reach still came as a surprise.

"They were building a new database," Ben continued. "Every person saved by Dr. Richter's cure was being catalogued, their DNA analyzed and stored. The facility's director, Dr. Rashid Patel, was a Phoenix Order sleeper agent for fifteen years."

"What was the purpose of this collection?" the prosecutor asked, though her grim expression suggested she already knew.

"Preparation for Phoenix 2.0." Ben pulled up intercepted communications. "They accepted that the first attempt failed, but they were planning a more targeted approach. Instead of killing billions at once, they would identify and eliminate specific bloodlines over generations. A slow-motion genocide hidden as natural death rates."

The tribunal judges exchanged concerned glances. Judge Katherine Morrison, the panel's senior member, leaned forward. "How many such facilities have you discovered?"

"Seventeen confirmed, with another thirty-two under investigation." Ben's jaw tightened. "For every Phoenix Order member we publicly try, two more operate in shadow. They've adapted and become more careful. The trials have taught them to hide better."

From the defendant's section, Klaus Richter raised his shackled hand. He'd lost significant weight during his imprisonment, his once commanding presence reduced to hollow eyes and trembling hands.

"May I speak to this?" Klaus asked, his voice barely above a whisper.

Judge Morrison nodded reluctantly. Throughout the trials, Klaus's interventions had provided crucial context, though each revelation seemed to drain more life from him.

"Dr. Patel was recruited by me personally," Klaus admitted. "The WHO infiltration was part of a twenty-year plan. We didn't just want to control who died—we wanted to control who lived. The genetic database would let us shape human evolution subtly, over centuries."

"And you're revealing this now, why?" the prosecutor pressed.

Klaus's gaze found Ben. "Because my daughter's sacrifice means nothing if the Phoenix ideology survives in shadow. Every hidden cell I expose honors her memory and inches me toward... not redemption, that's impossible. But perhaps understanding."

Ben's secure phone vibrated with an urgent message. Sarah Chen's encryption signature appeared with a single word: "Bangkok. Now."

"Your Honors," Ben stood. "I apologize, but an operational emergency requires my immediate attention."

Judge Morrison frowned. "Mr. Carter, this tribunal—"

"Has my complete cooperation and all relevant evidence." Ben was already moving. "But Phoenix Order operations don't pause for legal proceedings. Lives are at stake."

He left amid protests from the defense attorneys, who claimed that his actions were theatrical grandstanding, following him into the corridor. Let them complain. Every minute spent in courtroom debate was a minute the Phoenix Order used to regroup.

In the secure anteroom, Ben activated his encrypted connection to Sarah. Her face appeared, tense with urgency.

"We've intercepted communications about a major Phoenix cell in Bangkok. They're not developing new weapons—they're recruiting. Ben, they're targeting the children of deceased members, indoctrinating the next generation."

The implications hit like a physical blow. The Phoenix Order wasn't just trying to survive—they were ensuring their ideology would outlive the current generation.

"How many children?"

"At least two hundred across seven countries. They're calling it the Academy of Human Excellence. On paper, it's an elite boarding school for gifted students. In reality..."

"It's a training ground for future genocides." Ben was already heading for the exit. "Assemble a team. We shut this down tonight."

"Ben, wait." Sarah's tone made him pause. "There's something else. Klaus Richter's daughter, not Eva, is a younger daughter from an affair. She's seventeen, and she's at the Bangkok facility."

The world tilted slightly. Eva had a sister she'd never known about. Another Richter child raised in the Phoenix Order's shadow.

"What's her name?"

"Sophia. Sophia Richter. And according to intercepted communications, she's being groomed as Eva's replacement—the new scientific genius who'll complete what her sister tried to stop."

Ben closed his eyes, feeling Eva's presence like a ghost at his shoulder. Would the Richter family legacy be salvation or damnation? The answer lay in Bangkok.

"I'll be there in six hours," he told Sarah. "And Sarah? Don't tell Klaus. Not yet."

The flight to Bangkok passed in a blur of intelligence reports and tactical planning. The Academy occupied a renovated colonial mansion on the city's outskirts, surrounded by walls that suggested prison more than school. Thermal imaging revealed sixty-three heat signatures—children aged twelve to eighteen, along with adult handlers.

"Rules of engagement?" his team leader asked as they approached the drop zone.

"No harm to children," Ben stated firmly. "They're victims, not enemies. Adult Phoenix members are valid targets, but we prioritize the extraction of minors."

"And Sophia Richter?"

Ben checked Eva's photo on his phone, wondering if her unknown sister shared those same intelligent eyes. "We save her. From them, and if necessary, from herself."

The assault began at 0200 hours. Ben's team moved with surgical precision, neutralizing perimeter guards silently. The Academy's security was significant but focused on keeping students in rather than operators out—a fatal oversight.

Ben breached the main building alongside Sarah, their movements synchronized from years of partnership. The interior was surreal—propaganda posters featuring evolutionary imagery, classrooms with genetics equations scrawled across blackboards, and dormitories segregated by "genetic potential."

"Laboratory wing," Sarah whispered, checking her scanner. "Multiple heat signatures, including one isolated in what appears to be a clean room."

They found Sophia Richter working alone in a state-of-the-art laboratory, her concentration absolute as she manipulated viral cultures. The family resemblance was unmistakable—Eva's bone structure, Klaus's intense focus. But something else too, a hardness that spoke of different choices made.

"Sophia Richter," Ben announced, weapon lowered but ready. "We're here to help."

The girl—young woman, really—turned slowly. Her eyes held none of Eva's warmth, only cold calculation.

"You're Benjamin Carter," she said, no surprise in her voice. "My sister's lover. The one who let her die."

The words cut deeper than any blade. "She chose to sacrifice herself. I tried—"

"You failed." Sophia returned to her work with disturbing calm. "But it doesn't matter. Her counter-virus was flawed anyway. Mine will be better."

"Yours?" Sarah moved closer, noting the viral samples. "Sophia, what are you doing?"

"What Eva should have done." The girl's hands never paused in their delicate work. "Creating a virus that enhances rather than destroys. Selective evolution without the genocide. The Phoenix Order's vision refined to perfection."

Ben felt history was preparing to repeat itself. Another brilliant Richter woman, another laboratory, another choice between humanity's future paths.

"Your father's vision led to millions of deaths," he said carefully. "Eva died, stopping it."

"Because they were crude, obvious." Sophia sealed a sample container with practiced ease. "Killing billions attracts opposition. But what if people chose enhancement? What if we offered superiority rather than imposed selection?"

"The Phoenix Order tried enhancement. Colonel Steiner and others—"

"Primitive attempts." Sophia's laugh was eerily like Eva's, but twisted into something darker. "I've moved beyond their crude methods. Would you like to see?"

Before they could respond, she injected herself with a prepared syringe. Sarah raised her weapon, but Ben stopped her. Whatever Sophia had done, shooting her wouldn't reverse it.

"Fifteen seconds," Sophia counted calmly. "Then you'll understand why Eva's sacrifice was ultimately pointless."

The changes began subtly—her pupils dilating, skin taking on a healthier glow. Then, more dramatically, as she moved with sudden fluid grace, her reflexes visibly enhanced.

"Improved cognition, enhanced physical capabilities, optimized immune system," she listed. "All from a single injection. No death, no genocide. Just evolution offered freely."

"At what cost?" Ben demanded. "What aren't you telling us?"

Sophia's enhanced eyes found his, and for a moment, he saw Eva's compassion break through. "The cost is choice. Those who refuse enhancement will be left behind economically, socially, and evolutionarily. Natural selection through market forces rather than viral weapons."

"That's still genocide," Sarah said. "Just slower."

"That's progress," Sophia countered. "Eva saved the weak to die naturally. I'm offering them transformation. Which is truly more humane?"

Explosions rocked the building as Ben's team encountered heavier resistance. Adult Phoenix Order members were fighting desperately to protect their indoctrination center.

"We need to go," Sarah urged. "Sophia, come with us. You don't have to follow your father's path."

"I'm not." Sophia gathered her research materials with enhanced speed. "I'm creating my own. The Phoenix Order dies tonight—you've seen to that. But from its ashes rises something new. Not an organization but an idea. Enhancement available to all, creating a humanity that can face tomorrow's challenges."

"Eva believed humanity could face those challenges as it is," Ben said.

"Eva was naive." Sophia's words carried surprising sadness. "Brilliant, compassionate, but naive. Climate change, resource depletion, cosmic threats—baseline humanity can't survive what's coming. I can give our species the tools to endure."

Another explosion, closer. The extraction window was closing.

"Choose quickly," Sophia said. "Take me prisoner and waste my gifts in a cell. Or let me continue Eva's work properly—saving humanity through transformation rather than preservation."

Ben saw the trap perfectly. Arresting Sophia would make her a martyr to the remnants of the Phoenix Order. Releasing her would unleash unknown consequences. Eva had faced a similar choice with the counter-virus—act with incomplete knowledge or let millions die.

"We take her," he decided. "The world deserves to hear her ideas in open debate, not secret laboratories."

Sophia smiled, and it was purely Klaus. "You think tribunals and trials matter? The future doesn't happen in courtrooms, Mr. Carter. It happens in laboratories, in choices made by individuals who dare greatness."

"Your sister dared greatness. It killed her."

"Yes," Sophia agreed as Sarah zip-tied her hands. "But it also saved everyone. Imagine what she could have accomplished if she'd dared to improve rather than merely preserve."

They evacuated through smoke and gunfire, Ben's team extracting sixty-three indoctrinated children alongside Sophia. The Academy burned behind them, another Phoenix Order facility reduced to ash. But Sophia's words haunted the victory.

Back in Geneva, Ben delivered Sophia to Coalition custody. She went willingly, almost eagerly, as if imprisonment was just another phase of her plan. Her enhanced capabilities were undeniable—she processed information faster, moved with uncanny precision, and showed no signs of the injection's strain.

"She's not wrong," Dr. Brunner said after examining Sophia. "Her enhancement virus is remarkable. No lethality, no targeting specific populations. Just... improvement. The ethical implications are staggering."

Ben stood in the observation room, watching Sophia work with the provided non-biological materials. However, she still found ways to advance her theories. Eva's unknown sister, carrying forward a twisted version of her legacy.

His secure phone chimed with a message from the prison. Klaus had been informed about Sophia and requested to speak with Ben. Against his better judgment, Ben agreed.

The video link showed Klaus in his cell, aged beyond his years. "She survived then. My secret shame, my hidden daughter."

"You knew she was at the Academy?"

"I suspected." Klaus's voice carried infinite weariness. "Her mother was a Phoenix Order researcher who died when Sophia was three. I ensured she received the best education and the most resources. I thought I was protecting her from my world."

"You were preparing her to inherit it."

"Perhaps. Parents tell themselves comfortable lies." Klaus studied Ben through the screen. "What will you do with her?"

"That's for the tribunals to decide."

"No." Klaus shook his head. "That's for you to decide. The tribunals deal with the past. Sophia represents the future—a future where Eva's cure is just the beginning. Will you let that future unfold?"

"Eva believed in humanity as it is."

"Eva believed in choice," Klaus corrected. "She chose to save everyone rather than select who lived. But what if everyone could choose to become more? Isn't that the ultimate expression of her philosophy?"

Ben ended the call without answering. Outside, Geneva carried on with its daily life, unaware that in a secure facility, Eva's sister was designing humanity's next chapter. The Phoenix Order was dying, but its questions remained: Should humanity evolve? Who decides? What price are we willing to pay?

Three months after Eva's death, the immediate crisis had passed. The cure was globally distributed, the trials were proceeding, and the conspirators were facing justice. But the deeper reckoning had just begun.

In his apartment that night, Ben held Eva's picture, wondering what she'd think of Sophia. Would she see a kindred spirit twisted by circumstance? A threat to the humanity she'd died protecting? Or something more complex—a continuation of their mother's work, evolution through choice rather than chance?

The Phoenix Order had asked who deserved to live.

Eva had answered: everyone.

Now, Sophia asked who chose to evolve.

And Ben feared that answer might be: everyone.

The reckoning wasn't just with the past—it was with the future the Richter family had made possible.

# Chapter 37: The Foundation

The glass and steel edifice of the Eva Richter Foundation rose from the shores of Lake Geneva like a monument to hope. Six months after her death, Ben Carter stood in what would be the main laboratory, watching technicians install the final pieces of equipment. Every biosafety cabinet, every genetic sequencer, every containment unit had been chosen with a dual purpose: to develop defenses against bioweapons and to ensure those defenses could never become weapons themselves.

"Mr. Carter," Dr. Marie Brunner approached, her weathered face showing cautious optimism. "The Seoul team just confirmed they'll participate. That makes forty-three international laboratories committed to the Foundation's protocols."

Ben nodded, his attention caught by workers mounting Eva's portrait in the main atrium. They'd chosen a photo from her university days—bright-eyed, smiling, unaware of the destiny awaiting her. It was how he wanted the world to remember her: brilliant and hopeful, before the weight of salvation settled on her shoulders.

"Any word from the Beijing facility?" he asked.

"Dr. Chen Wei sends his regrets. The Chinese government is still... cautious about international biodefense cooperation." Brunner's diplomatic phrasing couldn't hide the frustration. Some nations still viewed biological research through the lens of weapons rather than shields.

"He'll come around," Ben said. "When the next threat emerges—and it will—they'll need us."

His phone buzzed with a familiar alert. Klaus Richter was calling from his maximum-security cell, their weekly conversation that had become an unexpected cornerstone of the Foundation's work. Ben accepted the video call, and Klaus's gaunt face appeared on the screen.

"I've completed the analysis of the Mumbai samples," Klaus began without preamble. Prison had stripped away his social niceties, leaving only raw intellect focused on atonement. "The Phoenix Order variant they were developing has interesting properties. Not lethal, but designed to increase susceptibility to conventional diseases."

"A primer virus," Ben understood immediately. "Make populations vulnerable, then let nature take its course."

"Elegant in its cruelty," Klaus agreed. "I've developed seventeen different counter-sequences. The data is uploaded to the secure server."

It was their strange partnership—the architect of genocide providing intelligence to prevent future genocides. Klaus's life sentence meant nothing to him beyond time to undo fractions of the harm he'd caused. Every Phoenix Order secret revealed, every bioweapon neutralized, was another small payment on an unpayable debt.

"There's something else," Klaus continued. "Sophia sent a message through her lawyer. She wants to contribute to the Foundation."

Ben's jaw tightened. Eva's sister remained in Coalition custody, her enhancement virus under intense study. In six months, she'd shown no adverse effects from her self-modification—if anything, her capabilities continued to improve.

"Absolutely not," Ben said.

"She's developed a detection system for synthetic pathogens. Something that could identify bioweapons before release." Klaus's expression remained neutral, but Ben caught the father's hope beneath the prisoner's facade. "Whatever else she is, Sophia has Eva's brilliance."

"And your ambition. It's a dangerous combination."

"Yes," Klaus agreed quietly. "But perhaps necessary. The Phoenix Order was just the beginning, Ben. Other groups will try similar atrocities. The Foundation needs every advantage."

Ben ended the call without responding. Through the laboratory's windows, Lake Geneva sparkled in afternoon sunlight. Somewhere beneath those waters lay the encrypted servers containing Phoenix Order data—terrorist networks, weapon designs, personnel files. Information too dangerous for any single nation but too valuable to destroy.

"Dr. Brunner," he called. "Schedule a senior staff meeting. We need to discuss the Sophia situation."

An hour later, the Foundation's leadership gathered in the secure conference room. Besides Ben and Brunner, there were Colonel James Mitchell from the Coalition's military advisory board, Dr. Yuki Tanaka representing the Pacific Research Alliance, and surprisingly, Sarah Chen, who had taken leave from the CIA to help establish the Foundation's intelligence network.

"Sophia Richter wants to contribute," Ben began, activating holographic displays showing her work. "Her synthetic pathogen detection system could identify bioweapons days or weeks before traditional methods."

"At what cost?" Sarah asked. "Every gift from a Richter comes with hidden prices."

"That's unfair," Dr. Tanaka interjected. "Eva gave freely—"

"Eva gave everything," Sarah corrected. "Including her life. Are we prepared for what Sophia might demand?"

Ben pulled up Sophia's latest research, which had been submitted through her legal team. The elegance was undeniable—she'd created algorithms that could identify artificial genetic sequences with 97% accuracy, distinguishing engineered pathogens from natural mutations.

"It's brilliant," Brunner admitted reluctantly. "With this system, we could have detected the Phoenix virus weeks before release."

"She's seventeen years old and enhanced with an untested virus," Colonel Mitchell reminded them. "Every dictator in history started as someone trying to improve the world. Do we really want to enable the next one?"

The debate continued for hours, touching on ethics, practicality, and the ghost that haunted every decision—what would Eva do? She'd believed in redemption, in the possibility of change. But she'd also died because she trusted too easily, loved too deeply.

"We accept the research but not the researcher," Ben finally decided. "Sophia's work can contribute from behind Coalition walls. When she's eighteen, when she's shown years of stability, we can reconsider."

The compromise satisfied no one completely, exactly what Eva would have called a good solution.

That evening, Ben returned to his new routine. Three days a week, he taught at the University of Geneva, his Bioethics and Global Security course drawing students from around the world. Tonight's lecture covered the rise and fall of the Phoenix Order, utilizing declassified materials to illustrate how scientific brilliance could be harnessed for monstrous ends.

"The lesson isn't that science is dangerous," he told the packed auditorium. "It's that science without ethics is catastrophic. Every discovery can heal or harm. The choice lies not in the knowledge but in how we apply it."

A student raised her hand. "Professor Carter, what about enhancement? If we can improve humanity safely, don't we have an obligation to do so?"

The question hung in the air—Sophia's question, the Phoenix Order's question stripped of its genocidal intent. Ben had wrestled with it for months.

"Define 'improve,'" he challenged. "Faster? Stronger? Smarter? By whose standards? The Phoenix Order believed that genetic purity was an improvement. They were wrong, but their core error wasn't in the science—it was in believing they had the right to choose for others."

"But we make choices for others constantly," another student argued. "Vaccines, public health measures, environmental regulations. Where's the line?"

"The line," Ben said carefully, "is between preventing harm and imposing change. Dr. Eva Richter's cure prevented death. It didn't alter humanity—it preserved it. That's the model the Foundation follows."

After class, Ben found a familiar figure waiting in his office. FBI Special Agent Diana Webb, Director Webb's daughter, who'd joined the Foundation's enforcement division.

"We have a problem," she said without preamble. "Three Foundation researchers in São Paulo received identical packages today. Each contained a vial of what appears to be Sophia's enhancement virus and a note: 'Evolution is choice. Choose wisely.'"

Ben's blood chilled. "Sophia's in maximum security. How—"

"That's what we're trying to determine. But Professor if her enhancement virus is in the wild..."

The implications cascaded through Ben's mind. Sophia's virus wasn't lethal, but it was transformative. If people began enhancing themselves outside controlled conditions, humanity would likely split into two distinct groups: the modified and the natural. The Phoenix Order's vision was achieved through voluntary adoption rather than forced selection.

"Lock down all Foundation facilities," Ben ordered. "No one touches those vials. And get me a secure line to Sophia's holding facility."

Thirty minutes later, Sophia's enhanced features appeared on his screen. Six months of captivity hadn't dimmed her intensity—if anything, she seemed more focused, more certain.

"Hello, Ben," she said pleasantly. "I assume you're calling about São Paulo."

"How?" he demanded. "You're under constant surveillance."

"I am," she agreed. "But ideas aren't so easily contained. I published my research months ago—encoded in my legal filings, hidden in academic language. Anyone with sufficient brilliance could reconstruct my work."

"You're starting an arms race. Enhanced versus natural, exactly what your father wanted."

"My father wanted to choose who lived. I'm letting everyone choose who they become." Sophia leaned forward, her enhanced eyes catching the light. "Eva saved humanity as it was. I'm offering what it could be. Why is your way right and mine wrong?"

"Because transformation under pressure isn't really a choice. When enhancement becomes necessary for competition, it's coercion with extra steps."

"Like education? Like healthcare? Like every advantage parents seek for their children?" Sophia smiled sadly. "Ben, the genie is out of the bottle. My virus exists.

The question now is whether responsible organizations like your Foundation will help manage the transition or futilely try to stop it."

She was right, and Ben hated her for it. Once knowledge existed, it couldn't be suppressed. The Foundation could either guide the development of enhancement technology or watch it unfold in the shadows.

"If we agree to study your work officially," he said slowly, "what guarantee do we have that you won't release more advanced versions?"

"None," Sophia admitted. "But consider this—would you rather have one Richter woman working within the system or inspiring a thousand copycats working outside it?"

The video ended, leaving Ben with an impossible choice. In trying to prevent the Phoenix Order's forced evolution, had they merely delayed a voluntary one? Was humanity destined to split regardless, just through market forces rather than viral weapons?

His phone rang. Dr. Brunner, her voice tight with urgency.

"Ben, you need to see this. Seoul just reported that three researchers requested the enhanced virus. Beijing had seven. Mumbai, five. They're not being coerced—they want the modifications."

Ben closed his eyes, seeing Eva's face, hearing her last words about everyone surviving together. But what if humanity no longer wanted to survive as it was? What if the Phoenix Order had been wrong about the method but right about direction?

"Staff meeting, one hour," he ordered. "Full leadership, maximum security protocols. And Dr. Brunner? Have our bioethics committee on standby. I think we're about to face our first real test."

The Foundation, which Eva's sacrifice had made possible, now faced a challenge she couldn't have imagined. Not how to stop a bioweapon, but whether to embrace one. Not how to preserve humanity, but whether to help it transform itself.

Outside his office window, Lake Geneva reflected the setting sun like molten gold. Somewhere beneath those waters, the Phoenix Order's darkest secrets lay encrypted. But the future's secrets were being written now, in laboratories where brilliant minds wrestled with questions Eva had died to keep closed.

The Eva Richter Foundation would honor her memory. But increasingly, Ben wondered if that meant preserving her vision or adapting it to a world that seemed determined to evolve with or without permission.

Six months after saving humanity, Eva's legacy faced its greatest test: what to do when humanity no longer wanted to stay saved.

# Chapter 38: The Price of Victory

The small Bavarian cemetery lay hushed beneath fresh December snow, exactly one year after Eva Richter had saved the world. Ben Carter's footsteps crunched along the gravel path, each step measured and deliberate, like a soldier approaching a position he'd defended at terrible cost. In his left hand, he carried white roses—her favorite. In his right pocket, pressed against his chest pocket, was the only photograph they'd ever taken together.

The grave was simple, as she would have wanted. A granite headstone bearing only her name, dates, and a single phrase in German: "Sie wählte alle"—She chose everyone. No mention of the Phoenix virus, the counter-cure, or the three million who'd died before her sacrifice took hold. History would remember those details. This place was for something more personal.

Ben knelt in the snow, his scars pulling tight in the cold. The one across his ribs from von Hess's blade. The bullet wounds from Colonel Steiner. Physical marks that would never fade, just like the deeper wounds that didn't show.

"Hello, love," he said quietly, placing the roses against the headstone. "I brought your favorites. The florist said white roses in December were impossible, but I insisted. Seems appropriate—you specialized in the impossible."

A year. Twelve months since she'd died in his arms, her body the final battlefield between extinction and survival. The world had moved on with shocking speed, as worlds do. The Phoenix Order trials had concluded with forty-one convictions. Klaus Richter served his life sentence by continuing to expose hidden cells and neutralize dormant weapons. The Eva Richter Foundation had grown from desperate hope to a global institution.

But here, in this quiet corner of Bavaria, time stood still.

Ben pulled out the photograph, edges worn from handling. Prague, three days before the end. They'd stopped at a Christmas market, momentarily forgetting they were hunted. An elderly woman with an ancient Polaroid had insisted on capturing them—the professor and the scientist, disheveled and exhausted but radiating something the woman had called "truth."

In the photo, Eva was laughing at something Ben had whispered. His arms encircled her from behind, his face buried in her hair. They looked like any couple in love, not fugitives racing against genocide. It was the only moment they'd stolen from fate.

"The Foundation's doing well," Ben continued, his breath misting in the cold air. "Forty-seven laboratories now, with more joining monthly. We've stopped three bioweapon attempts this year—copycats trying to recreate your father's work. Each time, your cure was there first."

He paused, wrestling with harder truths. "Sophia's been released. Turned eighteen last month, and the Coalition couldn't hold her without charges. She's established her own laboratory in Singapore, calling it the Prometheus Institute. Subtle, right? She's offering an enhancement to anyone who can afford it. The waiting list is three thousand names long."

The wind picked up, swirling snow around the headstone. Ben pulled his coat tighter, but didn't move to leave. These weekly visits were his anchor, the one place where he could speak honestly to the woman who had changed everything.

"I don't know what you'd think of her," he admitted. "Your sister has your brilliance but none of your wisdom. She sees humanity as a project to be optimized, not people to be saved. Last week, she published a paper arguing that refusing enhancement will become a form of child abuse within a generation. The ethics committees are in uproar."

His phone buzzed. Diana Webb, messaging about another Phoenix Order remnant located in Argentina. Ben silenced it. The work never ended, but it could wait an hour.

"Sometimes I wonder if we really won," he said to the stone. "Yes, humanity survived. Yes, the Phoenix Order failed. But the questions they raised won't die. Should we enhance ourselves? Who decides? What price is acceptable? Sophia says I'm a hypocrite—that I'll hunt bioterrorists to extinction while opposing bioenhancement that could save millions. Maybe she's right."

He placed the photograph against the headstone, next to the roses. "But you knew the answer, didn't you? It was never about being right. It was about choosing everyone, refusing to decide who deserved to live, evolve, or improve. That's why you tested the cure on yourself—not to prove you were right, but to prove you'd risk everything for everyone."

The cemetery's silence was broken by approaching footsteps. Ben turned, hand instinctively moving to the concealed weapon he always carried now. But it was just an elderly woman with a cane, moving carefully across the icy ground.

"Entschuldigung," she said in German, then switched to accented English. "You are the American? Who visits every week?"

Ben nodded warily.

"I am Greta Hoffman. I live in the village." She gestured toward the small cluster of houses beyond the cemetery. "I wanted to thank you."

"For what?"

"My grandson—he lives in Berlin. He was at the airport when the poison was released. The hospitals gave him the cure just in time. He lives because of her." She nodded toward Eva's grave. "Because of you both."

"I didn't—" Ben started to protest, but Greta raised her hand.

"The newspaper printed your picture during the trials. You held her while she died, yes? You loved her?"

"Yes," Ben said simply.

"Then you saved him, too. Love that inspires sacrifice—it multiplies. My grandson he studies medicine now. Says he wants to help people like the woman who helped him." Greta smiled, deep lines crinkling around her eyes. "One saved life creates more saved lives. This is how we defeat the darkness."

She placed her own flower—a simple alpine bloom—on Eva's grave and walked away, leaving Ben alone with a truth he'd been avoiding.

His phone buzzed again. This time, it was a message forwarded from the Foundation's education department. A student from his Bioethics course wrote to thank him for the semester.

"Professor Carter," the message read, "I wanted you to know that your class changed my perspective. I came in thinking neutrality was wisdom—that avoiding difficult choices meant avoiding mistakes. But you taught us about Dr. Richter, about how she chose to act even when action meant death. You showed us that standing aside while evil operates is just another form of collaboration. I'm changing my major to biodefense. Someone has to stand guard against the next Phoenix Order. Thank you for showing me that choosing sides isn't just acceptable—it's necessary. —Maria Santos"

Ben read the message twice, feeling something shift inside his chest. For a year, he'd carried Eva's death like a weight, focusing on what was lost. But Maria's words, like Greta's grandson, showed what had been gained. Not just lives saved, but minds changed, futures redirected.

"You clever woman," he said to the headstone. "You knew, didn't you? That saving everyone meant more than just keeping them alive. It meant showing them they were worth saving, that strangers would die for their right to exist unchanged. That's the real counter-virus—proof that love trumps selection."

He stood, knees protesting the cold ground. One more truth to share, the hardest one.

"I've met someone," he said quietly. "Dr. Sarah Nakamura joined the Foundation six months ago. Brilliant virologist, terrible cook, laughs at my worst jokes. She knows about you—everyone does. But she also knows that loving you doesn't mean I can't love again. I think... I think you'd like her."

The admission felt like betrayal and healing simultaneously. Eva had saved the world so life could continue. Hiding from that life dishonored her sacrifice.

"I'll always love you," Ben continued. "Every Phoenix Order cell I destroy, every bioweapon I help neutralize, every student I teach to choose action over apathy—it's all for you. But I'm going to try living too. Really living, not just surviving between missions."

He touched the headstone one last time, feeling the cold granite beneath his fingertips. Real, solid, permanent—like the change Eva had wrought in the world.

As he turned to leave, movement caught his eye. A young woman stood at the cemetery entrance, her features familiar despite the enhancement she'd undergone. Sophia Richter had come to visit her sister's grave.

"Ben," she acknowledged, maintaining careful distance. "I didn't expect anyone else to be here."

"It's Thursday," he said simply. His regular day, though he hadn't realized others had noticed.

Sophia approached slowly, her enhanced grace making her seem to glide across the snow. She carried her own flowers—yellow roses, the color of new beginnings.

"I never knew her," Sophia said, placing the flowers next to Ben's white ones. "Father kept us separated, different cities, different lives. I learned about her from news reports and trial transcripts. My sister, the savior I never met."

"She would have loved you," Ben said, surprising himself with the certainty. "Despite everything, she would have tried to save you, too."

"From what? My choices?" Sophia's enhanced eyes showed depths of emotion her controlled voice hid. "I'm not the enemy, Ben. I'm just someone asking different questions."

"The same questions your father asked."

"No." Sophia shook her head firmly. "He asked who deserved to live. I ask who chooses to change. The difference matters."

They stood in silence, two people shaped by Eva's life and death, representing the poles of humanity's future—preservation and transformation.

"The Foundation could use your detection systems," Ben said finally. "Official partnership, full oversight, ethical review boards. Work within the system instead of around it."

Sophia smiled slightly. "Sarah predicted you'd offer that. Yes, I've been in contact with Dr. Nakamura. Brilliant woman. You have good taste."

Ben ignored the personal observation. "And?"

"I'll consider it. But Ben, change is coming whether we guide it or not. The question isn't if humanity will enhance itself, but how. Wouldn't you rather have Eva's sister helping shape that future than strangers who never knew her?"

She didn't wait for an answer, walking away with the same enhanced grace. But she paused at the cemetery gate.

"For what it's worth," she called back, "I think she'd be proud, not of what you've preserved, but of what you've inspired. The students, the researchers, the survivors—they're her real legacy. We're just the guardians."

Alone again, Ben took one last look at Eva's grave. The white and yellow roses looked beautiful together against the snow, past and future, preservation, and change, two sisters who never met. Still, they would shape humanity's course for generations.

He pulled out his phone and texted Sarah Nakamura: "Dinner tonight? I have stories to tell."

Her response was immediate: "Always. Your place or mine?"

"Somewhere new," he typed. "Time to try something different."

As he walked back through the cemetery, Ben felt lighter than he had in months. The scars remained, would always remain. The work continued, and would always continue. But Eva hadn't died to create a world frozen in amber. She'd died to give humanity the choice to continue—messy, chaotic, beautiful in its diversity.

The price of victory was everything she'd given and everything he'd lost. But the reward was that everyone who lived because she'd chosen them all.

One year later, the world turned on. Enhanced and natural, preserved and transformed, all of humanity's varied paths spread out from a single moment of sacrifice. And somewhere in that spreading future, Eva Richter's love continued its work, not selecting who deserved tomorrow, but ensuring everyone had the chance to see it.

Ben Carter walked on, guardian of that chance, carrying her memory into whatever tomorrow would bring.

The price had been paid.

The victory, however complex, was real.

And in a small Bavarian cemetery, two kinds of roses bloomed against the snow, proving that even in ending, there could be a beginning.

# About the author

My career, spanning over four decades, has been a testament to the power of strategic vision and leadership. From the vibrant sales floors of Bashinski's Gems and Jewelry to the strategic boardrooms of Reeds Jewelers and Friedman's incorporated, I have navigated the intricate world of diamonds, gems, and the buying sector with a blend of scientific precision and creative flair. My passion for storytelling is not just a personal interest, but a reflection of my professional journey. It is beyond the sparkle of a well-cut diamond, weaving narratives that resonate with the heart and mind—my passion is clear in my published five nonfiction books and the twenty-plus novels. As an author, I understand the value of legacy, and it's the timeless beauty of a family heirloom or the enduring impact of a well-told tale. My books are more than just collections of words. They are vessels of 'knowledge, experience, and imagination' destined to inspire and enlighten. I hope you find these sources of information and entertainment too.

# Also by Donald J. Wright

**Novels**

Lilith's Garden

ASIN: B0DQX8ZWD9

The Terraforming Protocol ASIN: B0FHBVY1QS

ASIN: B0DNY8Z3WB

The Prometheus Protocol

ASIN: B0DLHFF79M

13$^{th}$ Moon Book I

ASIN: B0DGNTV533

13 Moons: Legacy of the Guardians Book II

ASIN: B0FDYNP7WP

Killer Ice

ASIN: B0F1G6HVMR

The Ghost Code

ASIN: B0F4FGQMG5

The Golden Book

ASIN: B0DXQGMFL8

Tomorrow

ASIN: B0FFTS4C39

The God Equation

ASIN: B0FGZFNZTD

THE QUANTUM SCHISM:
ASIN: B0D1N9RHMQ
The Quantum Alchemist:
ASIN: B0FD43QCDB
The Quantum Heart:
ASIN: B0F9YZTRVG
The Codex Protocol:
ASIN: B0F1Z1XH89
THE QUANTUM ECHO
ASIN: B0F6KWPGG2


**Non-Fiction**
Beyond Climate Debates
ASIN: B0DZB8CB7K
Diamonds Under Fire
ASIN: B0CDYSTBLL
The Handbook of Lab-Created Diamonds
ASIN: B0D8V4X3CW
The Diamond Revolution
ASIN: B0FHBVY1QS
Eternal Shine
ASIN: B0DQX8ZWD9
Globe Treasure Hunting
ASIN: B0DF6RN4H8

www.ingramcontent.com/pod-product-compliance
Lightning Source LLC
Chambersburg PA
CBHW031214020726
47499CB00002B/574